The Steam Engine in Industry
I The Public Services

Historic Industrial Scenes

THE STEAM ENGINE IN INDUSTRY ~1

the public services

George Watkins

MOORLAND PUBLISHING

British Library Cataloguing in Publication Data

The steam engine in industry.
  1: The public services. - (Historic industrial
  scenes).
  1. Steam-engines - Great Britain - History -
  Pictorial works
  I. Watkins, George   II. Series
  621.1'0941   TJ461

ISBN 0-903485-65-6

ISBN 0 903485 65 6

**Printed in Great Britain
for the Publishers**
Moorland Publishing Co Ltd,
9-11 Station Street, Ashbourne,
Derbyshire, DE6 1DE, England

Contents

Acknowledgements

Every reader will realise, from the wide range of places recorded, how completely I have depended upon the goodwill of the owners of the plants and their officers. It was their courtesy, ready help and so often, enthusiasm, which alone made my record, now unique, possible, and with their willingness to grant facilities, so easy. There were so many people that I can only thank them all for their assistance in recording the great days of steam. I cannot, however, omit a special tribute to Mr. F. E. Durham and the mechanical engineer's staff of the Metropolitan Water Board. With them, as in other fields, my relations became very much personal, and more than ever welcome now that so much has been lost.

Bath University again has provided many facilities, friendly interest and encouragement to continue my work. The privilege of a Visiting Fellowship has been especially valuable in retaining so many links. I am also indebted to my good friends Tony and Jane Woolrich who, providing quiet and uninterrupted facilities for study and writing, greatly assisted the selection and arrangement of the examples and descriptive matter.

My debt to Colin Wilson is deep indeed — his many weekends of artistry are evident in the delightful plates which bring out so much detail from negatives which were often basically bad, or have deteriorated over the years. The sources of my information acquired as the folly of a misspent life were many, including the *Engineer; Engineering; Pumps and Pumping Machinery*, J. Colyer, 2 vols: 1892 & 1899; *Pumping Machinery*, H. Davey, 2nd Edition, 1905; *Pumping Machinery*, A. M. Greene, New York, 1919; *Modern Pumping and Hydraulic Machinery*, E. Butler, 2nd Edition 1922.

Introduction

The steam engine brought great benefit to mankind and it is fairly safe to state that the tremendous advances of the last two centuries would not have occurred without it. Although one of man's inventions that was largely beneficial, it did create overcrowding in towns near to the new industries and the results caused grave concern. At the same time, however, the steam engine provided the remedies with pure water supplies and by removing sewage. The overcrowding resulted in epidemic diseases, distress and high death rates, and these epidemics spread far beyond their point of origin. Inspectors were appointed to investigate conditions in those areas where deaths from epidemic disease exceeded 23 per 1000 annually. Watford was an example, which indicates that the bad conditions known to exist in the large towns and cities were by no means confined to them, and many small communities instituted remedies, once the cause was established. 'Watford is in a very filthy condition indeed, and the existing arrangements are quite insufficient to remedy the evils', reported the inspector in 1849. The townspeople of Watford spurred on by the ever-present fear of cholera, were among the first to adopt the Public Health Act of 1849 and create a local Board of Health. The first problem to be dealt with was sewage disposal, and when this was overcome, they turned to water supply, and began a scheme costing £5,000 in 1854. This embraced a 100,000 gallon reservoir and 6 inch bore supply main, and a pair of horizontal steam engines.

The plant used for water supply was as varied as the nature of the demand. During the latter part of the nineteenth century many country houses were supplied by pumps driven by the Davey Safety Motor using steam at 2 to 3 psi, and needing little attention, other than refilling with coke every few hours. The requirements of small towns were met by simple slide-valve engines, economically pumping water to meet the day load and to a reservoir that stored a supply to meet the reduced night load without pumping. The large towns with a heavy demand required plant to work continuously for months on end without stopping. In view of such constant operation the highest economy was demanded and contracts to supply such engines were won or lost upon consumption guarantees of a fraction of a pound of steam per pump horse-power hour.

Steam power was equally valuable in the fens areas, since there the first efforts in drainage in the seventeenth century led to land shrinkage that became an ever increasing problem. Thus Thomas Neate writing in 1748 on the flooded state of the Manea district, noted that there were 250 windmills draining the Middle level, and 50 in the Whittelsey area alone. Despite such concentration of windmills flooding was still a serious handicap to farming until the coming of steam power, which was able to operate whenever it was needed. For this purpose the engines had to be rugged, reliable and simple to keep in order.

The steam engine, with good designers, was able to meet these conflicting needs, and the aim of the book is to illustrate a few of the delightful machines which did so much for us all, and to indicate how widely the influence was exerted. The plan has been to provide a short introduction outlining the general background of the use of the steam engines in the several areas of public service followed by plates of the various types, each with a brief descriptive note. I may have seemed prolix in describing the background of the need for and use of the engines — my justification is that it gives the whys and wherefores of my subjects, and adds interest by the variety of engine and plant designs developed to solve the problems. My aim has been to illustrate the widest range of types and makes of plant involved, and I have expressly avoided railway, traction, and fairground engines since these are fully covered already.

In addition to water supply, sewage disposal and land drainage I have included the use of steam power for electricity and gas supply, leisure use in small ships, and a few other sundry uses in the public services.

Consideration of the technical aspects of the steam engine have likewise had to be omitted, but my books *The Stationary Steam Engine* and *The Textile Mill Engine*, and also *The Industrial Archaeology of the Steam Engine* by R. H. Buchanan and G. Watkins, contain limited information upon this aspect. F. Woodall's *Steam Engines and Waterwheels* is also informative upon the subject.

Although most of the engines illustrated have been scrapped, several fine examples are preserved by responsible authorities and great efforts by local groups, and run under steam periodically. The majesty of steam will thus be preserved, a great credit to all those concerned.

Technical Notes

Pumps

My examples indicate the importance of pumping in the public services, and the following notes explain the terms used in the descriptions.

The ram or plunger pump (Fig. 1) comprised of a circular ram or plunger working within a chamber. Fluid was drawn in through the inlet valve on the upward stroke, to be discharged through the outlet valve on the downward stroke. These pumps were vertical or horizontal as desired.

The double-acting bucket pump (Fig. 2) consisted of a bucket working closely within a machined bore. Inlet valves allowed the fluid to enter, to be discharged through the outlet valves on the return stroke. The single-acting bucket pump was similar to the double-acting but was usually vertical, and fitted with valves in the pump bucket. The action was that fluid again was drawn in through inlet valves and passed through the valves in the bucket on the return stroke. As the bucket ceased moving downwards the valves closed and the fluid was drawn up with the bucket on the next upward stroke, and delivered through outlet valves.

The bucket and plunger (Fig. 3) combined (Fig. 1) and (Fig. 2), drawing fluid through the inlet valve on the upward stroke, and discharging on the downward one through the outlet valve. Some of the fluid however passed to the space above the bucket, ie around the plunger, to be discharged on the next upward stroke. This gave a double-acting effect with a single set of valves.

Centrifugal pumps (Fig. 4) operated entirely by rotary action in which a rotor fitted with curved vanes spun the fluid outwards within a closed casing. The rotor was mounted upon the driving shaft, which was horizontal or vertical, and driven directly or through gearing, from the engine (Fig 5).

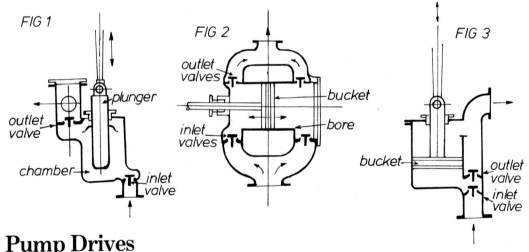

Pump Drives

Bucket or plunger pumps could be driven from any point with a reciprocating motion, as the composite drawing indicates.

In a beam engine (Fig. 6), they were driven: from the ends of the beam (either directly or by a piston tail rod); from an intermediate point; or by a separate crankshaft through gearing. As a rule the well lift and the surface forcing lifts were by separate pumps, each from its own rod. Occasionally, as at Liverpool (Plate 4), the separate pumps were in line and driven from a single point. The pump driving-rods were either attractive cruciform castings (eg Plate 58), or neat wrought iron or steel forgings. Timber rods with iron strapping were uncommon, except for horizontal drives to well pumps, for which they were flexible yet stiff.

Horizontal engines drove by a tail rod (Fig. 7), or by connecting rods and quadrants to the pump rods. Inverted vertical engines (Fig. 8) drove the force pumps by side rods passing around the crankshaft to the pumps below, and well pumps from a separate crankshaft.

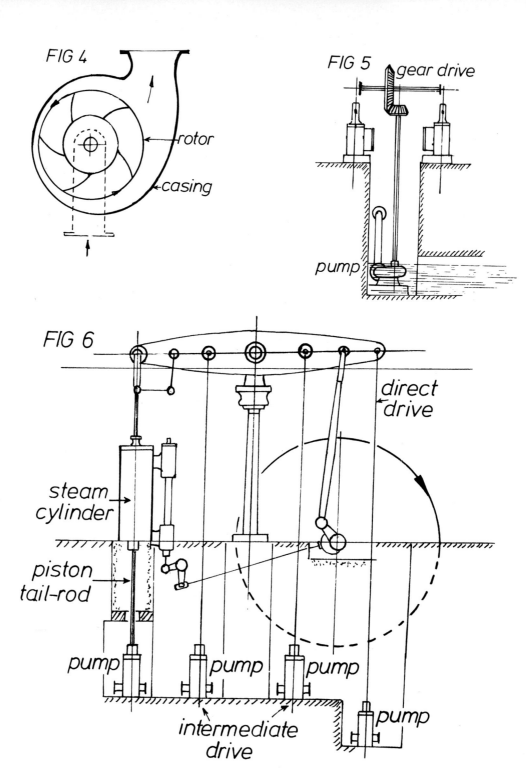

FIG 4

rotor

casing

FIG 5   gear drive

pump

FIG 6

direct
drive

steam
cylinder

piston
tail-rod

pump   pump   pump

intermediate
drive

pump

8

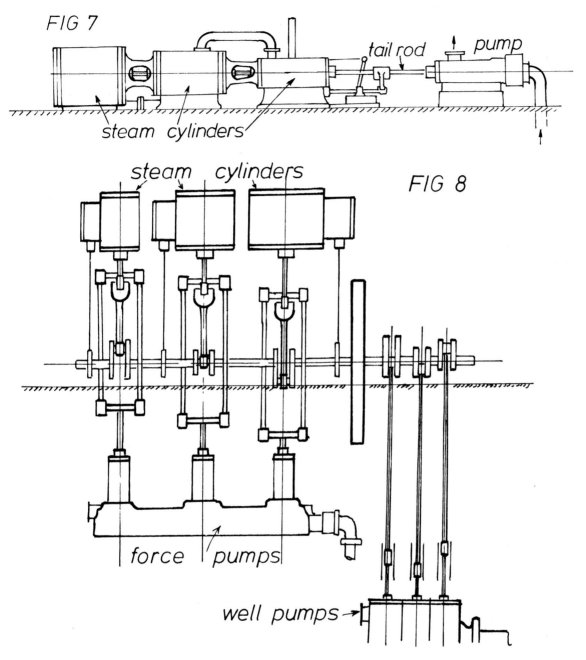

FIG 7

tail rod   pump

steam cylinders

steam cylinders   FIG 8

force pumps

well pumps→

The Cornish Steam Cycle

Most of the engines which are illustrated worked upon the double-acting cycle in which steam is alternately admitted to, and exhausted from, each end of the working cylinder during each revolution. This was patented by James Watt in 1782, and prior to this, steam had only operated upon the piston on the downward stroke. However, the single-acting cycle was still much used for pumping, especially for those engines which worked on the 'Cornish cycle'. The sequence of events was as follows (Fig.9):

A Engine at rest, with the exhaust valve open, to be followed by the opening of the steam valve, when the combined effort overcoming the resistance of the pumps will drive the piston downwards.
B At about one third of the stroke the steam valve closes to complete the stroke by the vacuum through the exhaust valve and expansion of the steam above the piston.
C Engine has completed the downward stroke and all valves are closed, and the engine pauses until
D The equilibrium valve is opened to allow the steam above the piston to pass to the underside, when the weight of the pump rods (or bob-weight) draws the piston to the top again, to pause until the cataract control opens the exhaust and then the steam valves to begin another cycle.

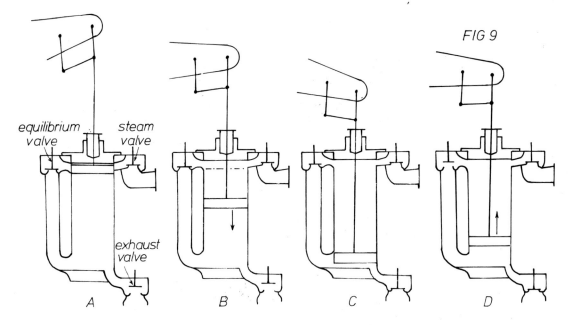

FIG 9

Glossary of Technical Terms and Abbreviations

hp power developed in horsepower
rpm revolutions per minute
spm strokes per minute
psi steam pressure in lbs per square inch
gph gallons per hour
mgd million gallons per day
head the total resistance created by pumping, including lifting from well to the surface, forcing to reservoirs, and the frictional resistance caused by the mains, the whole being expressed in the equivalent of feet raised vertically
lift height the water was raised to ground level
force height water was forced above ground level to reservoir
duty the designed capacity of one engine
no of engines number of units that type and size on the site
date in brackets at the right-hand side the year the photograph was taken
engine dimensions bore of the engine cylinder in inches and the piston stroke in feet and inches
rotative engines those in which the engine and pump were connected with a rotating crankshaft and flywheel system
direct acting steam and pump cylinders directly connected without rotating motion

Water Supply

The need for a supply of pure water to a large population with a limited supply was felt early in London, and first met there by the waterwheel-driven pumps of Peter Moris in 1582, and Sir Hugh Myddleton's New River Scheme. Steam power was also tried there almost as soon as it was practicable and Thomas Savery installed one of his engines at York Buildings in 1713–14. It was not really successful and was followed by a Newcomen engine which, starting work in 1726, was stopped in 1731 owing to the high cost of the fuel consumed. The need remained however, and an improved Newcomen engine was installed on the site in 1752, and was soon followed by another one, and together these ran for twenty years.

The first Watt engine with a separate condenser for public water supply was that erected at Shadwell, London in 1778, and another was installed at York Buildings in London in 1810. Other engines followed as water supply companies evolved, but the great impetus in this development stemmed from the disease epidemics caused by the intake of water from the polluted London section of the River Thames. These led to the prohibition of water collection from the tidal section liable to sewage pollution, the intakes then being removed to new pumping stations upstream operated by each of the several water-supply companies. It was this situation which gave London its great series of engines, totalling over 500 when the Metropolitan Water Board was established in 1902. River water is usually pumped by low-lift units into settling basins for purification and storage, with high-lift pumps to deliver it to the service reservoirs for local use. As other, and higher level areas were developed, repumping units became necessary to supply them and London eventually had a number of these, often to supplement supplies from earlier engines on local wells.

I have used London as a general introduction to waterworks practise since only there were all of the methods to meet the needs of a large local population adopted. The fact that much of this occurred at the hands of separate companies, gave the engines delightful individuality at the choice of their several engineers.

Town water supply is mainly from wells or rivers, and the source determined the machinery used. Well water was usually pure enough for direct use, to pump straight into the mains and reservoirs, but in some areas treatment was needed to reduce the hardness. Usually thirty to fifty per cent of the power was required to raise the water to a small surface balance reservoir, and the notes describe some of the methods used. The surface lift to the high level reservoirs was usually by pumps driven directly from the engine beam, the crossheads, or in horizontal engines, the piston tail rod, or later, through gearing from the crankshaft.

River supplies needed two stages of pumping, firstly by massive low-head pumps to settling reservoirs and filters, and then, when purified, by high-head pumps into the town service mains and reservoirs. The pump loads were quite different, since in wet weather the low-head collecting pumps would work at full load to fill the settling basins while the water was available, and do little when the river flow was reduced. The high level, pure-water pumps would work at a nearly constant load to meet daily consumption with economy.

The large authorities used splendid stately engines running 10 to 20 rpm, with the pumps connected directly to them. Smaller areas, with the need to reduce initial costs, often used faster engines driving the pumps by gearing, which reduced the cost of the house and engines. Easton and Amos especially made many such small engines, and these served local authorities everywhere.

In view of the importance of water supply a brief note upon some of the builders of pumping engines is justified. Each achieved worldwide repute for the reliability and economy of their products, usually with designs in which they specialised, yet often differing greatly from those of their rivals.

Harvey and Co of Hayle in Cornwall were engineers of the highest standing before they supplied the first Cornish engine for waterworks service, for Old Ford, East London in 1838. The high economy soon led to installations in a number of London pumping stations, and the business justified the retention of a large staff of drivers, mechanics, even painters, to maintain it. A number of Bull engines were also installed. The Kew Bridge plant remains to show the superb quality of their work when encouraged by enlightened authorities. Their last engine for London was probably the rotative triple-expansion engine of 1894 at Ferry Lane, East London (Plate 46) and the beam engine for Wansunt, Kent of 1903. The works were closed in 1904.

Easton and Amos, later Easton and Anderson; Easton Anderson and Goolden; Easton and Co; and from 1908 a branch of The Pulsometer Co of Reading, served the water supply industry for nearly seventy years of the nineteenth century. They certainly made a wide variety of steam engines for pumping services. The early ones were attractive overbeam grasshopper engines (see Plate 78), and then 'A' frame (Plate 21) and 4- and 6-column beam engines. For low-lift pumping, the design range was great (Plates 78 to 82). They usually used slide valves even for the largest, and for triple-expansion engines, and their last large engines were triples (similar to Plate 48) for Antwerp, and other cities, at the end of the nineteenth century.

Hathorn Davey and Co, Sun Foundry, Leeds took over the business of Carrett Marshall and Co. on the same site in 1871. Henry Davey of Tavistock was the leading light and a brilliant engineer who, from 1870 developed the differential steam control system. His compound and triple direct-acting pumping engine was used worldwide for mines and waterworks. The last engine of this type — a compound built in 1923 — was for a Far East tin mine. From 1895 they developed the triple-expansion rotative engine to high economy with a very simple and durable valve gear, making the last of these about 1935. They have long made a wide range of pumps for all services, as, now owned by Sulzers, they continue to do.

James Simpson and Co started in business in 1790 and were established in Pimlico by 1859. Long specialising in economical pumping engines, they also made engines for many other services, including breweries, flour mills, and marine service. They took the UK licence for the Worthington engine in 1886 and rapidly developed its use to supply over 750 engines in all by 1950 for almost every pumping service, except possibly land drainage and dry docks. The virtues and simplicity of this design gained them the great Coolgardie Water Supply contract for twenty-two engines. It was the only type which could meet the stringent conditions that was imposed for economy and time of delivery. The works were moved to Newark in 1900, and like Hathorn Davey their later engines were inverted vertical triple-expansion rotative engines of the highest economy. The last triple-expansion engine contract was completed, for Dover in 1951, and long specialists in all types of pumping machinery, they are still fully active in this and other fields.

The large concerns were equally at home with plant for the smaller authorities, whose needs were also met by many builders of smaller and simpler engines, some of whom, such as Tangyes, Hayward Tyler and Co, Joseph Evans, Camerons-Pearns, either specialised in pumping machinery or included it in their range of products. The engine builders of the Eastern Counties also contributed, with pumping machinery by Marshall, Robey, and Dodman, concerns better known for their work in other fields.

The large textile-mill engine builders also played a part, and of them, Foster; Yates and Thom; Galloway; Fairbairn Lawson Combe and Barbour, all made pumping machinery. Ashton Frost and Co of Blackburn were highly successful in this field, and, from the building of mill engines (mostly up to 500 hp for the weaving trade), began making pumping machinery of high quality early in the twentieth century — the two engines for Lincoln's Elkesley pumping station gained the highest premium for economy of fuel in 1911. Douglas and Grant, Fleming and Ferguson, Barclays and John Cochrane in Scotland, among others also built fine waterworks pumping machinery.

Cornish Engines

The Cornish engine was the highest development of the first successful steam engine of 1712, which evolved into the most impressive of all steam engines with the great Dutch drainage units of the 1840s. It comprised an overhead rocking beam with the pumps placed at one end, driven by a steam cylinder working upon the other end. It was a brilliant conception that allowed variation in the proportion and placement of the parts, and one which persisted for over 250 years. There was no crank or flywheel, the motion being controlled entirely by the action of the steam valves. The form did not allow of great variations of the general layout.

1   The South Staffordshire Waterworks Co, Moors Gorse Pumping Station   (1930)
James Watt & Co, Birmingham, 1875. 1½mgd to 500ft. 20psi. 65in x 10ft. Cornish Valves, 8spm. Pumps 22 and 21in x 9ft.

These were interesting as waterworks engines since both sets of pumps were placed outside of the engine room. They lifted about 180ft from wells at the back of the house, by bucket pumps which were driven by an auxialliary beam placed above, and driven by links from, the main beams, which themselves were continued through to drive the surface lift pumps at the front. The two main engines were placed at 12ft 6in centres, and the main beams were 33ft long and 6ft deep. Watts' engines were usually plain with little wasted upon ornaments which paid nothing, and the fluted steam trunks, the cast-iron panels over the valve chests, and the neat timber casing around the cylinder, were specified by the South Staffordshire Co, who desired that the engines should be neat as well as reliable. They were replaced by electrical pumps after eighty years of service, and the fine conditions of the engines was a tribute to the company and their staff. There was a similar pair of engines at the Huntington station. The handwheel seen upon the left-hand fluted trunk controlled the speed at which the engine made the equilibrium stroke.

2   Derby Waterworks, Little Eaton Station. Two Engines   (1932)
R. & W. Hawthorn, Newcastle-upon-Tyne, 1849. 1.2mgd to 170ft 48in x 8ft. Cornish Valves. 10spm, 40hp. Plunger pump 19in x 8ft 0in.

Derby was early in providing a public water supply, and this plant illustrates the concern that it should be attractive. Supplying the low-service area, it followed the current Gothic revival tradition in the decoration of the structure supporting the beam centre bearings. The pump bob-weights seen on either side in the foreground show the same care for detail, with a fine fluted mid-section, neat top and bottom flanges and the hemispherical top, all in cast iron, and well set off by the neat guard rails around them. Synchronising gear was fitted to ensure alternate working when the two were in use at the same time. They gave little trouble in eighty years of service with no major incidents on record, a tribute to all concerned in the building and running since the engines pumped directly into the main and reservoirs without a stand pipe to maintain a constant head. If a main burst the sudden loss of head was very dangerous in a Cornish engine if the engineman did not act at once to stop it. It is probable that the four Cornish boilers were the original ones as the steam pressure was low, but despite this, the plant was quite efficient for a small unit. They were followed in 1873 by a pair of Kitson rotative beam engines for the high-service system, which were also in the Gothic tradition. All was scrapped about 1937 following the installation of electric pumps.

3   Liverpool Waterworks, Dudlow Lane Station. One engine   (1936)
Rothwell & Co, Union Foundry Bolton, 1869. 1mgd from well to reservoir. 56in x 10ft. 8–10spm. 30psi.

The Liverpool Waterworks engines were names after various officers on the Board and this was *T. Duncan*. Steam had long been a standby to electric pumps at this station due to the growth of the load, but the steam plant was kept in working order until scrapping at an unknown date. The bucket pumps lifted the water about 230ft from the well, and the ram forced it to the reservoirs about 180ft higher than the station. The twin fluted columns and the circular motif in the top framing were neat, and an attractive feature was the Liver birds, Liverpool's civic symbol, cast on the entablature cross-beams above the columns. The parallel-motion links were between the beam webs, with Gothic shaped flanges on the brasses. Two Lancashire boilers with superheaters supplied the steam for this, and an adjacent rotative engine.

1

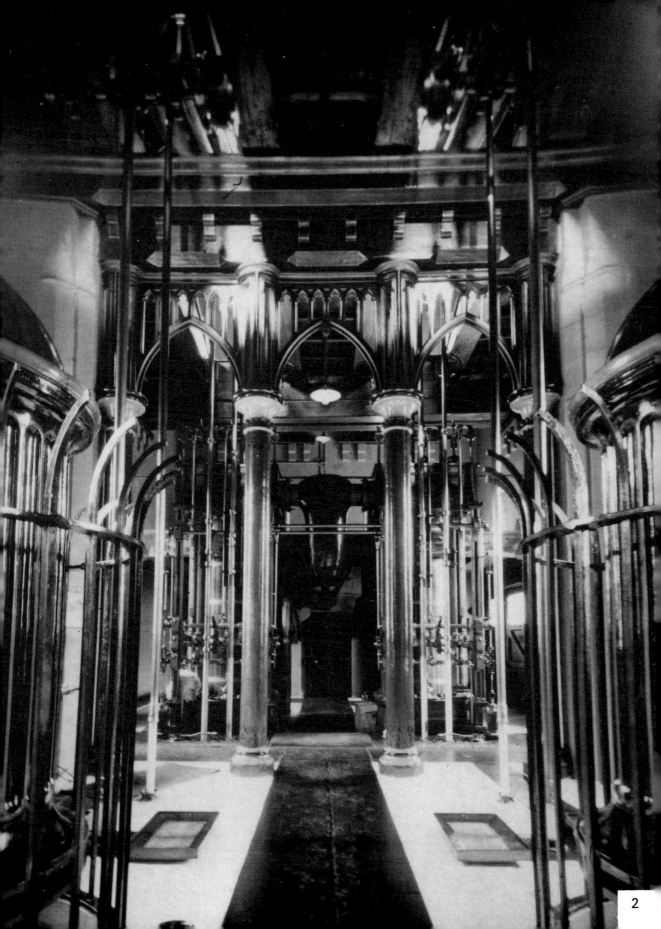

2

3

4   Liverpool Waterworks, Green Lane Station. The Holmes Engine   (1936)
Harvey & Co, Hayle, 1845. abt 1mgd to 180ft. 20psi. 50in x 10ft. Cornish valves. 8–10spm, 69hp. Well and surface pumps — sizes unknown.

This was a good example of Harvey's waterworks engines when they were becoming a regular part of the works output. Little was spent on ornament in this engine, yet it was attractive with a neat casing over the cylinder covers, and the valve chest with moulded panels on the front. The four columns were slight, and together with neat hand railing gave the engine a quiet dignity that was highly attractive. The pumps seen at the lower left-hand corner show how a Cornish engine was adapted for driving well and surface force pumps. The pump bob-weight was again a fluted casting with neat flanges, and the force pump itself was supported upon girders in the well below the chequered floor plates. The well pump, many feet below, was driven by side rods which passed around the force pump from the crosshead above the bob weight. In an adjacent house to the right was another slightly larger Cornish engine named *George Holt* which was made by G. Forrester of Liverpool in 1851. This had the well and steam cylinder placed the opposite way to *Holmes*.

5   The beam loft of the *Holmes* engine was a pleasing place. The beam webs again were generally plain, with the simple circular boss, but neatly ribbed around the edges; a rare feature was that the makers' name and address and date were enclosed within a shield-shaped ornament, in contrast to the usual plain lettering cast on the beam. The railings, protecting the sweep of the catch wings which restricted the movement of the beam in the event of loss of load, were the finest I met for this purpose. The parallel motion was of the usual high standards of the Cornish blacksmiths' craft, well finished and with delightful keeps above the cotters which adjusted the bearing brasses. At this period, Harveys sometimes used a cast-iron crosshead and Cornish 'C' collar for the piston rod fixing, but here they adopted a forged iron crosshead with a cotter and a thread and nut. The gap above the centre bearing brasses was customary in the Cornish engine which had no upward loading, in fact, many such engines ran with no top keeps at all.

6   Most of the Cornish engine beams were plain open sand castings with the parallel motion between, or outside the webs. A small number, however, were of the open or lattice type, and this is an example as fitted to two of the Cornish engines, (Nos 4 and 5) at the Hammersmith Station of the old West Middlesex Water Co. This was another delightful piece of Cornish craftsmanship, some 30ft between the end centres and 5ft deep. A number of the London Cornish engines were fitted with top tension stays after cracks developed in the beam of one engine. The vertical lattice pillars were guards for the sweep of the catch wings, which restricted the movement if the beam travelled too far.

7   The valve gear was always a delightful feature of Cornish engines and, beautifully finished, together with the size and stately movements of the engines led to the view that they were the finest engines made by man. There were usually three arbors or shafts, each controlling the valve for one part of the cycle, ie, steam, equilibrium and exhaust. The curved levers, and fine forging and finish, were the peak of the engine smiths' art, and were enhanced by years of loving care from the engine men. This, on a London waterworks, was typical of all of their splendid engines, as those preserved at Kew Bridge Pumping Station illustrate.

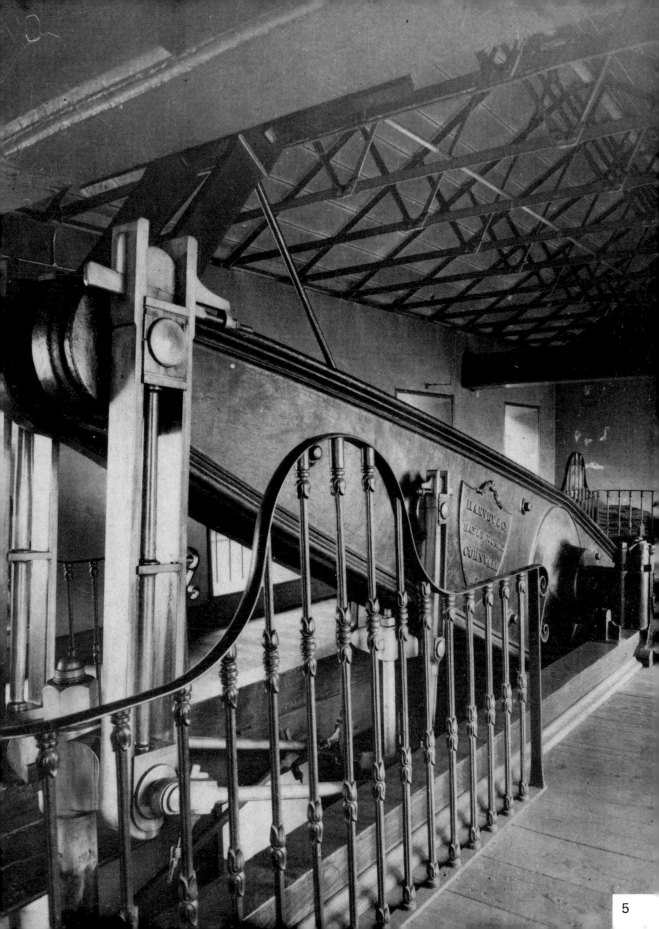

5

Bull Engines

This was a Cornish engine with the cylinder inverted, and was named from the engineer who used this design in the 1790s. The pumps were placed below the cylinder and were driven by the piston rod which projected through the lower cover, which reduced the size and the cost of the engine house, since there was no overhead beam. Although it operated on the Cornish cycle of storing energy in a massive moving weight, and with the Cornish steam cycle, it was not as economical as the beam type. Even so it was adopted in the late nineteenth century, at the Severn Tunnel, where two engines of 50in cylinder bore were erected in 1876 (see Plate 100).

8   The Metropolitan Water Board, Campden Hill Station. Three engines   (1932)
Nos 1 & 2 Engines: Harvey & Co, Hayle, 1857. 5½mgd to 136ft. 50psi 71in x 10ft 0in. 8–10spm, 55hp Ram pumps 32in x 10ft 0in. No 3 Engine: Harvey & Co, 1873. 90in x 10ft. Ram pump 40in x 10ft 0in. 8¾mgd to 136ft.

These again showed Cornish engineering at its best, and the clean design, high finish and neat but unobtrusive ornamentation were notable. The skill of the smiths' and fitters' work in the valve gear indicates the quality of Harvey's work when their London business was at its height, and they retained a team of craftsmen there for maintaining the many Cornish engines supplying water to the Metropolis. The finely shaped false covers (two were over 8ft and the other 9ft 6in diameter) were as attractive as the rest of the engines. The plug rods in front which operated the valve gear were guided by the inverted plunger cases seen at the top floor level. These were a peak of engine building, working with no noise other than that of the valve gear, which together with the air pump for the condenser, were driven from a small beam below the engine room floor, coupled to the piston rod.

9   Scarborough Waterworks, Osgodby Station. One engine   (1938)
Kitson & Co, Leeds. abt 1mgd to 350ft? 50psi? 45in x 9ft. Cornish valves. 8–10spm.

This worked well and surface force-lift pumps from the piston rod, the pumps being in line below, with the well set driven by bridle rods passing around the ram surface-lift set, placed just below the engine. Although an unpretentious engine, it gave very good service for many years until it was replaced by electrical pumps in the 1930s. It was however retained as a standby engine, and in 1939 as a precaution against electricity failures during the war, it was good enough to justify an advertisement for a driver in the local press. Interesting variations from the Harvey engines (Plate 8) were the adoption of twin plug rods to drive the valve gear (although Harvey also did this) and the placement of the equilibrium valve at the top of the cylinder, driven by a rod from the top arbor of the valve gear. Another interesting feature was the use of a diaphragm-type expansion joint in the equilibrium steam trunk; this can be seen at the top of the circular trunk or pipe at the left-hand side of the valve gear.

frontis

Birmingham Waterworks, Whitacre Station   (1937)
James Watt & Co, 1885. abt 5 mgd. Head 250ft. 80psi. 33 and 60in x 10ft 0in. 8spm. Ram pumps 26in x 10ft.

These splendid examples of civic engines showed what James Watt & Co could do when an enlightened civic authority wanted plant that was attractive as well as economical. By no means all engines working with a simple steam cylinder operating a pump below were Bull engines, and these were an example of Henry Davey's final development of that type, and were in fact, of his differential design. The steam cylinders were compound and double acting, with their motion coupled by a system of linking beams which allowed each side to assist the other, and to keep them in phase to give even water flow. They were economical, and with steam cut-off at 0.5 of the stroke in the high-pressure cylinder, gave eight-fold expansion of the steam. Although highly ornate, the decorative features were economical in that the fluted columns, the eagle brackets supporting the top platforms and the motif around the lower platform were all repetitive. Even the holding-down bolts had ornate covers. They were 26ft high to the top of the cylinders.

8

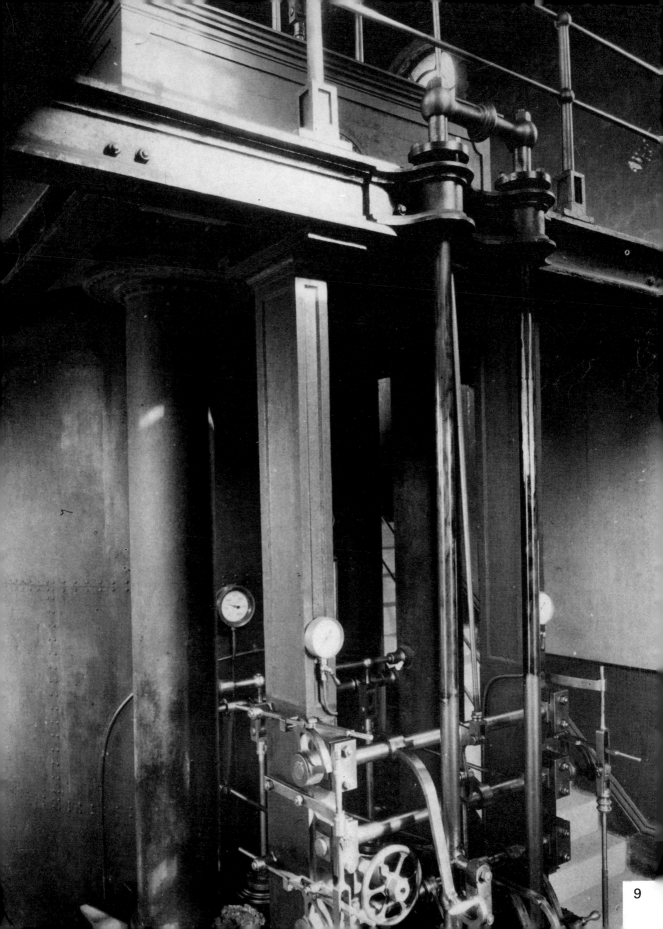

9

Direct-Acting (Non-Rotative) Pumps

The simplest mechanical pump had the steam and water cylinders coupled together on a single rod, but it was inefficient since little steam expansion was possible. This could be remedied if the steam was expanded in two or three cylinders in succession, or if the steam and water ends were coupled together through a varying leverage system, or by hydraulic compensators in which an excess of effort at the beginning of a stroke was stored to be returned at the end.

10   Nottingham Waterworks, Boughton Station. One engine   (1957)
Hathorn Davey and Co, Leeds, Wks No 3440, 1881. 1½mgd. 400ft. 45psi. 24in and 44in x 6ft slide valves. 8–10spm. 2 x 20in bucket pumps in well.

This was almost certainly the sinking pump for the Papplewick engines, and removed for similar service at Boughton. It was retained as the well pump for the first Boughton plant, to serve the high lift triple-expansion Fairbairn Lawson Combe and Barbour engine. Originally supplied with a condenser and quadrants, the latter were altered for the Boughton wells by substituting 'T' bobs for the original 'L' type. It was 65ft long over the pump railings to the end of the condenser, and pumped up to 2mgd. It was Davey's standard design of the period, with domed cylinder covers and twin low-pressure piston rods passing beside the high-pressure cylinder to the crosshead.

11   Wolverhampton Waterworks, Cosford Station. Two engines Nos 4 and 5   (1934)
Hathorn Davey and Co, Leeds, Wks No 4616, 1888. 3mgd. 440ft. 30psi. 2 cylinders, 56in x 5ft to each. Slide valves. Well pumps 27½in x 4ft 6in.

These worked upon the principle that if the steam and pump connecting rods are coupled at an angle to each other, but to the same pin, the ratio of relative movement gives a mechanical advantage through the stroke, and permits expansive working. Thus at the start of the stroke when the steam pressure is at a maximum the steam piston moves more slowly than the pump, but as the stroke proceeds the steam piston moves faster, while the pump rod, as the pin is moving over to top, travels more slowly. These engines replaced a pair of Cornish beam engines working on steam at 30psi, and were made twin simple-expansion to use the same boilers, which then were fairly new. One engine drove the well pumps, and the other the forcing set to the high-level system, and the balance went to a low service. The well engine (No 4) was later converted to compound working.

12   Metropolitan Water Board, Hammersmith Station. One engine   (1953)
James Simpson and Co, London, Wks No 2382, 1890. 5½mgd. 195ft. 80psi. 22in and 44in x 4ft. 20dspm. 230hp. Bucket pumps, 19½in diameter x 4ft.

This was a high development of the Worthington engine embracing the oscillating cylinder compensating system of J.D. Davies (1879) as improved by Worthington and adopted for waterworks in 1885. It comprised two oscillating cylinders 8in diameter mounted upon trunnions, with plungers working upon a crosshead between the steam and pump ends. They were oil filled and connected to a pressure storage tank. As the stroke progressed they opposed the effort of the steam cylinders up to mid-stroke, pumping the oil into the storage pressure tank, and after mid-stroke, the compensators then returned the effort to the crosshead to allow completion of the stroke with the steam cut-off. This engine was especially designed to replace two Cornish engines in the same space.

13   Wisbech Waterworks, Marham Station   (1937)
Worthington Simpson & Co, Newark, Wks No 3107-8, 1936-8, 2mgd. 275ft. 180psi. 12, 19 and 30in x 2ft. Semi rotary valves. 33dspm. 195ihp, 175phP. 15½in pump rams.

These were standard Worthington engines, in which two worked together and side by side with one engine controlling the steam valves of the other side. In this way neither side could move until the other one had made its full stroke, to re-open the valves to give steam to the other. Economy was secured by expanding the steam in three cylinders in succession. The ratios were 113 sq in in the high pressure to 706 sq in for the low-pressure cylinders and by cutting off at 0.75 of the stroke in the first cylinder 8½ expansions and 16.75lb of steam per pump horse-power per hour was achieved. There was a single semi-rotary slide valve below each cylinder, together with an auxiliary cut-off valve for the high pressure one, driven through a lost-motion system to allow the valves to remain open for most of the stroke, and the engine to come to rest easily at the end of the stroke.

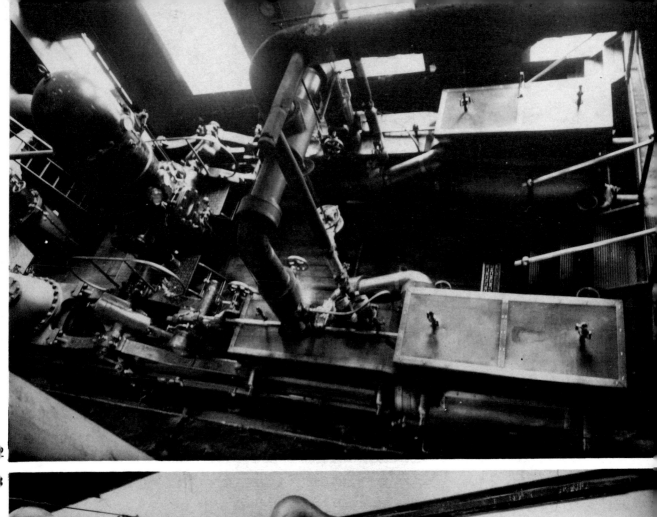

House-built Single-cylinder Rotative Beam Engines

In this type the piston was attached to one end of the beam, and the other end was coupled to a connecting rod, crank, and flywheel, the whole being built into the house. It had two advantages over, and needed less attention than, the Cornish type since (a) the stroke was controlled by the crank and flywheel system, and (b) expansive working was achieved by the energy stored in the flywheel. In my examples the pumps were driven directly by rods coupled to the beam.

14   Bristol Waterworks. Chelvey Station. Two engines   (1932)
Robert H. Daglish, St Helens Foundry, 1866. abt 1½mgd. 300ft. 60psi. 31½in x 6ft 6in. Slide valves. 14–16rpm.

These were disused in the 1930s, although still usable as standby units. Nothing was available regarding the pumps, but they were probably double-acting bucket type, pumping from a culvert to the high-level service reservoirs near Bristol. The cylinders were fitted with Meyer cut-off valves, with Bristow's anti-friction device to relieve the pressure on the backs, and were Daglish's No 4 application of the type. The beams had twin cast-iron webs, and were 21ft long, with enlarged circular ends, and 3ft 4in deep in the middle, were neatly moulded outside. The connecting rods worked between the webs, with the pumps driven by the double channel-section rods close to the connecting rods. The beam centre was 18ft above the floor, and the flywheel 18ft in diameter was fairly massive, but the crank, carrying little load was slight, only 3½in thick. The middle or packing platform was 8ft above the floor, and two engines were placed at 12ft 6in centres. An interesting feature was that the radius rods of the parallel motion were very short.

15   South Staffordshire Waterworks, Hopwas Station. Two engines   (1937)
Gimson and Co, Leicester, 1880. 1mgd. 200ft? 35psi. 24in x 4ft 6in. Slide valves. 20rpm. 75hp each.

These were built for the Tamworth Waterworks, and called *Spruce* and *Woody* after council members. Plain and neat, they had double-web beams, with the unusual feature that, since the pumps were on the same line as the top connecting-rod centres, the pump rods were continued around the crankshaft by a cast-iron bridle, and a similar one outside of the crank to make the pump rods identical. The cylinders were cased in attractive timber strips, and were 6ft 6in high, with the valve chests and Meyer adjustable cut-off valves at the front. The beams were about 14ft long, and the flywheel approximately 13ft 6in diameter was made in four pieces, with eight arms. Steam was supplied by a reducing valve from two Cornish boilers which also steamed the Tangye horizontal-tandem compound engine which drive well and surface pumps by gearing in the next house.

16   Hinckley Waterworks, Snarestone Station. Two engines   (1937)
Bever Dorling and Co, Dewsbury, 1892. 50psi. 24in x 4ft 6in. Slide valves. 30hp (designed). 20rpm. 0.5mgd. 320ft?

In contrast to the previous example this neat pair of engines had a single decorative column beneath the beam centre-bearings, and were fitted with drop valves placed at the back of the cylinder. The neatness was increased by the narrow timber bands covering the cylinder lagging, and polished cast-iron false top covers. They were built to pump 12 gallons per stroke, later increased to 20gps, making the power up to 50hp. The well pumps were at the end of the beams, beyond the crank, and the force or surface lift was between the connecting rod and the beam centre. Like Plate 10, the railings had cast-iron columns and polished steel handrails, the columns being attractive for the polished pointed finials. The well pumps were driven directly from the gudgeons in the beam webs, but the force pumps were driven through a parallel linkage to give vertical motion. Three Cornish boilers by Spurr, Inman supplied the steam, latterly with two on to drive the increased load on the engine in use.

14

16

Woolf Compound House-Built Beam Engines

In this type, the two cylinders, one high pressure and the other low pressure, were placed close together at one end of the beam, and along its centre line. There was often very little steam-receiver space between them, and the high-pressure cylinder was placed nearest to the beam centre bearing. It was therefore of shorter stroke, and was often set upon a stool to keep the top cylinder covers level. The cylinders were occasionally placed side by side, ie across the beam, and so were of equal stroke, as with four of the old Kent Water Co engines of the 1890s.

17   Leicester Waterworks, Cropston Station. Two engines   (1936)
Neilson Bros, Glasgow, 1870. 1.7mgd. 135ft. 30psi. 15in x 4ft 6in and 32in x 6ft 0in. Drop valves. abt 60hp. 15rpm. 1 pump no details.

Contrasting in most ways with Plate 18, these were fine examples of Messrs Hawksley's practice as consulting engineers, in combination with a responsible authority. The engine house matched the rest, as the stairway to the upper floors was placed in a turret at the corner of the engine room outside. The pump was driven from the beam by the single rod seen near to the connecting rod, each having a neat forged centre boss. The low-level crankshaft and valve driving-gear below the floor added to the general attractiveness which, with the decorative centre cross structure and the high finish of the valve gear represented the best tradition of Victorian craftsmanship. The timber casing around the steam pipe at the back of the engine room was a sheer delight. With growing demand for water the Easton and Anderson triple-expansion engine had carried the load, but the wrought iron boilers of the Neilson engines were reinsured in 1939 for their original pressure when 70 years old.

18   Metropolitan Water Board, Crayford Station, Kent. No 3 Well   (1934)
Harvey & Co, Hayle, 1883. 2mgd. 260ft. 110hp. 26in x 5ft and 42in x 7ft 6in. Drop valves. 12rpm. Well pump 37in x 4ft. 3 force pumps 18in x 3ft.

There were three rotative engines at Crayford all made by Harveys: a horizontal 24in x 5ft (1860), a single-cylinder rotative (1874) and No 3. The photograph is actually of No 2, as it shows the well pump better; No 3 had a cast-iron disc crank but otherwise was similar outside, and was in the house at the right. The pumps were unusual in that the crankshaft and flywheel, and the well pump were in the open air, and the force pumps, driven by a three-throw directly coupled crankshaft, were in the glass fronted house. When built, No 2 had Cornish-type tappet valve gear but this was later altered to the rotating shaft type with which No 3 was built. No 3 was a pure Woolf compound with three valves only at the top and bottom, so that the steam passed directly from the high to the low pressure cylinder by a single valve.

19   Wallasey Waterworks, Liscard station, Wallasey. Two engines   (1937)
Fawcett, Preston and Co, Liverpool, 1894. 1.4mgd. 260ft. 80psi. 24in x 4ft 11in and 35in x 6ft 0in. Slide valves. 10-12rpm. 1 bucket and plunger pump.

These were massive engines with a single pump of 16in and 25in diameter x 4ft stroke placed 160ft down the well. The single web beam was 25ft end centres, and the pump was driven by the twin timber rods with the cross braces seen near to the crank. The cylinders were cased in felt covered with hard wood, and were each fitted with a Meyer valve which, driven by two eccentrics gave variable cut-off to each cylinder. The flywheels, 23ft in diameter, with eight circular arms, had some hollow rim sections to balance the weight of the crank and connecting rod. The cylinders, columns and crankshaft bearings were all set upon heavy cast-iron girders in the floor, and the crank end of the beam was made with a jaw end and pin upon which the single top-end bearing of the connecting rod worked. The piston rods were guided by crossheads working upon circular guide bars, another plain feature, simpler than the usual parallel motion. There was also a horizontal tandem-compound Fawcett engine of 1902 with cylinders 35 and 60in bore in regular use in 1937. Three boilers supplied the steam for both beam and horizontal engines.

17

18

FAWCETT PRESTON & Cº Lᵀᴰ
1894
ENGINEERS LIVERPOOL

19

Independent Rotative Beam Engines

In this type, the beam centre bearings were supported by a framework stiff enough to resist most of the stresses at that point, with none, or a minimum, carried by spring beams along, or an entablature across the engine house. Space will only allow illustrations of a few designs for such structures.

20   Tunbridge Wells Waterworks, Pembury Station. Two engines   (1936)
James Simpson and Co, Pimlico, 1866. 0.6mgd. 370ft. 30psi. 30in x 3ft 6in. Slide valves. 25rpm. 50hp. 1 bucket and plunger pump.

This very neat design was adopted by James Simpson for small pumping engines in the mid-nineteenth century. The single pump was placed near to the crank below the strong box bed, and was driven directly from the beam by the two stiff circular rods, seen just behind the crank, joining at a crosshead on the top of the pump plunger. An interesting feature was that two bolts only held the crankshaft bearings to the bed by collars, the bolts then continued upwards to hold the top keeps. The cast-iron beams were 11ft end centres and the flywheel the same in diameter. They were compounded by adding a high-pressure cylinder 16in diameter below and tandem to the original one in 1891.

21   Metropolitan Water Board, Brixton Hill Station. Two engines   (1936)
Easton and Anderson, London and Erith, 1875. 2.25mgd. 230ft. 130hp. 60psi. 21in and 42in x 5ft 0in. Slide valves. 22rpm. 130hp. 1 pump 21½in x 2ft 5¾in.

The Brixton station of the old Lambeth Water Co served higher level areas in South London, repumping water received from the western intake and filtering plants. It grew to nine beam engines by 1900, and these were replaced by diesel engine sets in 1937. Two of the beam engines were made by Easton and Anderson which, of the same design, differed in size, this one being slightly smaller. They were cross compounds, ie with a high- and low-pressure cylinder each driving to its own beam and crank. The single double-acting pump was near to the cylinder and bolted to the underside of the bed. Driven directly from each beam by a circular forged-iron rod in this way, only light stessing occurred in the framing. All of the motion work, and the double-webbed beams were also iron forgings. The cast-iron flywheels were 15ft in diameter, and weighed 7¾ tons, the beams were 16ft end centres and some 16ft above the bed.

22   The Chiltern Hills Water Co, Dancers End Station. One engine   (1935)
J. C. Kay and Co, Bury, 1867? ?mgd. 250ft. head. 50psi? 18in x 2ft 6in. Meyer slide valves. 14–18rpm. 1 well pump per engine.

The twin six-column engine was a good type for small waterworks, although rarely adopted for the purpose. It was an attractive type, strong, open and accessible, with a repetitive motif in the cast iron work that reduced the cost of columns and frames. When, as here, the pump was placed close to the cylinder, the stresses in the top were reduced leaving the framing to carry little but the equalising effect of the massive flywheel between the two sides. The wooden casing over the cylinder lagging was interesting, as there were no brass bands to keep it in place. The use of the forged-strap type of connecting rod ends was well adapted for a general shop where the blacksmith was a mainstay in construction work.

23   Scarborough Waterworks, Irton Station. Two engines   (1936)
Bradley and Craven, Wakefield, 1884. 1mgd. 300–500ft. 55psi. abt 28in x 5ft 6in. Drop valves. 16–18rpm. hp and pumps not known.

The central frames were 8ft square, and consisted of an 'A' shaped casting on each side, bolted to a rectangular box top section, with the beam centres 18ft above the floor. The light spring beams and entablature rested upon the top section. The single web beams were 22ft end centres by 3ft 6in deep, and the flywheels were rather small, ie 14ft in diameter. The crankshaft was below the floor and unusual in that it was double webbed to drive the low-lift pumps which, in the central well in the octagonal central railings, were driven by a disc crank on the outer end of the crankshaft. The force pump was driven from the beam by the parallel motion and rod near to the connecting rod. The wooden casing over the cylinder lagging was especially fine in these engines, as beside covering the cylinders and valve chests, it was also carried over the top cylinder and valve chest covers, which were usually open and polished. The connecting rods were delightful castings.

23

Rotative Beam Engines, Gear Drive

The cost of the large beam engine and its house weighed heavily upon the smaller authorities who strove to provide a clean piped water supply, and one solution was to drive the pumps by gearing from a small fast-running engine, which, often of an independent type was housed in the simplest structure. This also allowed the engine and pump to be placed to better advantage.

24 Southampton Waterworks, Timsbury Station. Two engines   (1935)
Easton and Anderson, London and Erith, 1879. 0.5mgd. Head unknown. 60psi. 18in x 2ft 0in and 24in x 3ft 0in. Slide valves. 45rpm. ?hp. Pumps 8½in x 2ft 0in.

Easton and Anderson played a large part in water supply engineering, and these were a good example of their practice for smaller works, being made for the old South Hants Water Co. There were two identical, but independent units, with an economical Woolf compound beam-engine driving its own pumps. The high-pressure cylinder was fitted with a Meyer cut-off valve. The cast-iron single web beam was 11ft between end centres, and 9ft 6in above the floor, and the six-arm flywheel, 14ft 6in diameter, was made in two pieces. The next engine at the station was that built by Bryan Donkin (Plate 26). The ever increasing demand outgrew these and the larger plant, and a Ruston oil engine and turbine pumps were installed in 1933, followed by another oil engine set in place of the Donkin in 1936. The little Easton engines were retained as standby through to 1950, when they were replaced by electrically-driven vertical-spindle pumps that pumped to the full head in one unit.

25 Southampton Waterworks, Timsbury Station. The pumps
Easton and Anderson, 1879. 8½in x 2ft.

The well and surface force-pumps, were similar in size and construction; each set, with its engine, being quite independent of the other. The central pair were in the well, and pumped the water to basins to settle the chalk by the Clark process, and the outer ones then pumped the softened water to the service reservoirs 12 miles away. The crankshafts were splendid examples of smiths' work, in wrought iron 9in diameter machined only for the pump-rod bearings, which had a collar forged on either side of the brasses. Each engine drove by a cast-iron pinion 3ft in diameter to a mortise gear 6ft in diameter upon the pump crankshaft. The teeth were 9in wide, those of the larger mortise wheel being made of apple wood, with twin tails passing through slots in the rim, and held in place by round pins driven through the tails. The teeth were lubricated with a mixture of blacklead and tallow, the central platforms giving access for this and to the oil holes in the top brasses of the pump connecting rods.

26 Southampton Waterworks, Timsbury Station. One engine
Bryan, Donkin and Co Ltd, Bermondsey, London, 1897. 1mgd. abt 200ft. 60psi. Cylinder sizes unknown. Slide valves.

Performing a similar service, but for double the quantity of water, the Donkin engine differed in almost every respect to the Easton engine in the station, except that it, too, was a Woolf compound. The beams were twin steel plates, with both the force pumps and the engine connecting-rods coupled by single bearings to pins between the webs. The twin sloping and fluted support columns for the beam centre were an unusual feature, and with the light upper framing, were most attractive, as was the oval section flywheel rim nearly 20ft in diameter. The entablature across and the spring beams along the top were built into the house, each adding to the appearance of lightness in the whole engine. The pumps too were very different, with the well set again outside the engine room, but driven by a pitch chain, although this could have been a later substitution for gearing. The two plunger force-pumps were worked by rods close to the beam centre.

24

Horizontal Rotative Engines, Direct Drive

Waterworks followed the general trend and adopted this type from the mid-nineteenth century (eg Great Yarmouth in 1855 and Crayford in 1860). My examples indicate how water authorities met their varying needs with units which, within a similar framework or structure, varied considerably in their details.

27   Chatham and Rochester Waterworks, Luton Station. One engine   (1957)
Marshall, Sons and Co, Gainsborough, No 52528, 1911. Duty unknown. 100psi. abt 24in x 3ft 6in. Drop valves. 20rpm. 2 bucket pumps abt 12in x 3ft 6in.

Water supply started from this station in 1866, probably with a Whitmore beam engine, to be followed by five others of which only this and a Simpson beam engine of 1901 remained. The Marshall engine was of their highly efficient type with drop inlet and exhaust valves driven by a rotating side-shaft, with governor, and with hand control for variable speed. There were two Glenfield and Kennedy double-acting pumps in-line driven from the piston tail-rod. The flywheel was 14ft 6in diameter, and with the pumps behind the engine was over 40ft long. The Simpson beam engine pumped 3½mgd to 380ft, the horizontal engine assisting local needs.

28   St Helens Waterworks, Eccleston Hill Station. One engine   (1935)
R. H. Daglish and Co, St Helens, 1897. abt 1mgd. Head unknown. 100psi. 18in and 32in x 3ft 6in. Slide valves. 20rpm? 100hp? Well and force pumps.

Water supply from this station started in 1853, with a Cornish engine, followed by another, each in separate houses. The Daglish engine was plain and named *Henry Marsden* with Meyer valves on both cylinders. The well pumps were driven by quadrants or 'L' bobs from the low-pressure piston tail-rod, with the force pump driven from the front quadrant. It had long been a standby set owing to increased water demand, but with the variable cut-offs it was still useful. The flywheel was about 14ft diameter.

29   The West Cheshire Water Board, Prenton Station. Two engines   (1959)
The Lilleshall Co, Oakengates, Shropshire, 1923. 2.4mgd. 490–600ft. 160psi. 26in and 54in x 4ft 0in. Drop valves. 20rpm. 300hp. Well and two sets of force pumps.

These were high-class engines designed for economy to lift from wells 280ft deep and force water to the Heswall and Prenton districts at different heads. The twin sets of force pumps were directly behind the low-pressure cylinders, on either side of the rod which drove to the well-pump quadrants, beyond the railings at the rear. The main engine frames were fine castings 20ft long, containing the bearings for the double-throw crankshaft. The flywheels were 21ft diameter, in two pieces. Steamed by Adamson boilers, they each developed 260hp in water lifted, using 8 tons of coal per day. All was scrapped upon electrification in 1961.

30   The South Essex Water Board, Linford Station. Two engines   (1966)
C. Markham and Co, Chesterfield, 1903. 50hp? 0.8mgd. 200ft. head. 120psi. 14in and 28in x 3ft 0in. 20rpm? Corliss valves.

Built by makers of colliery and ironworks engines, these were heavily built cross-compound sets, with a double ram-pump driven off the low-pressure tail rod. The condenser pumps were driven from the high-pressure tail rod, ie that near to the camera. The trunk frames and circular crosshead-guides were neat but heavy enough to transfer the effort necessary with the pump load on one side. They were designed to lift the water from shallow wells, but the level fell in later years, and air lift pumps driven from Ingersoll compressors were installed to raise the water to the surface. Increasing demand, and continued fall in the water level led to the installation of Allen vertical-spindle pumps driven by steam turbines in 1936, with pumps 130ft down the well, and forcing to 200ft head.

2

3

Horizontal Engines, Gear Drive

Horizontal engines developed rapidly to run at speeds of 80 to 120rpm, but the pumps worked most easily at 10 to 20rpm, and driving through gearing became widely used late in the nineteenth century. My examples illustrate this practice to give a quiet pump and efficient engine, although a little effect was lost in the gearing.

31   Leamington Spa Waterworks, Main Pumping Station. Two engines   (1956)
Young and Co, Eccleston Foundry, London, 1879. 1.4mgd. 190ft. 60psi. 21in x 3ft 6in. 40rpm. abt 70hp. 2 pumps 18in x 5ft, 9rpm.

These engines were named *The Wackrill* and *The Harding*, and arranged so that either engine could drive either set of pumps, but the couplings were later altered. There was a cast-iron curb around the well top 2ft 6in deep, with two massive girders across, to carry the pump crankshaft bearings. The gear wheels were 2ft 6in and 10ft diameter and ran quietly although entirely of cast-iron, and the pump crankshafts were later made to slide backwards to disengage when using the other pumps. Each engine drove two pumps by cranks at 180 degrees, giving a steady flow in the simplest manner.

32   Leamington Spa Waterworks. The Harding Engine

This was plain and substantial, with single strap-type ends to the forged iron connecting rods and four bars for the crosshead guides. The 12ft 6in diameter flywheel was in two parts with only two bolts at the hub, and internal spigot and cotters for the rim joint. An unusual feature was that Stephenson's link motion was adopted to give variable steam cut-off, adjusted by the lever in the sector. The pistons were originally fitted with tail rods, but these were later removed and new covers fitted. The surface condensers were placed in the water flow, with independent air pumps.

33   Stockport Waterworks, Wilmslow Station. One engine   (1958)
Marshall, Sons and Co, Gainsborough, 1895. abt 1mgd. ?ft Head. 100psi. 8in and 14in x 1ft 8in. Drop inlet and Corliss exhaust valves. 50hp. 135rpm.

The station was interesting in that the well and surface pumps were of differing types and makes (see also Plate 54). The Marshall engine drove the well pump through 5 to 1 reduction gearing and a wooden connecting rod to the pump quadrant near to the door. The large gearwheel was 5ft in diameter. The engine was Marshall's standard of the time, with the drop inlet-valves operated by Proell trip gear, and Corliss exhaust valves underneath operated by an eccentric on the crankshaft. The whole was very well laid out, with the piping beneath the floor, and the exhaust rising to the condenser and air pump at the back. The engine and condenser were nearly 25ft long overall and the pump quadrant was 20ft further on. The whole was neatly fitted at the side of the engine room, with the vertical force pumps on the right behind the railings. Two Tinker Shenton boilers of 1921 supplied steam up to 120psi. The whole was replaced by electrically driven turbine pumps in 1960.

34   The Great Western Railway Co, Kemble Station. Two engines   (1959)
The Hydraulic Engineering Co. Chester, 1903. 0.75mgd. 100psi. 18in and 32in x 2ft. 60rpm. 150hp. 2 Pumps abt 24in x 3ft.

The considerable water needs of a large works often cannot be met from the local council supplies, and natural local sources sometimes have undesirable features. This then leads to supplies being sought some distance away. This was the case with the Great Western Railway, and led to the utilisation of a supply available at Kemble Station on their line. Originally two Lancashire type boilers provided the steam, but these were later removed, after the installation of electrical vertical-spindle pumps and steam was then supplied as needed by a locomotive in the siding. It was an attractive plant with the two wells some 7ft 6in apart, and wooden sweep rods for the pump drives, 21ft long and 11 x 9in middle section. They were stiffened by metal straps top and bottom. The flywheels were 9ft in diameter, and the large gear wheels 12ft in diameter.

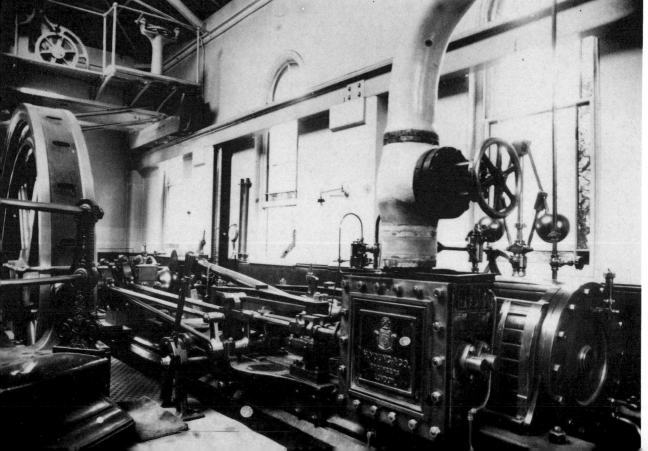

33

35   Metropolitan Water Board, West Wickham Station, Kent. One engine   (1952)
John Penn and Son, Greenwich, 1898. 1mgd. 200ft. 100psi. 12in and 21in x 2ft. Meyer slide valves.
abt 80rpm. 50–80hp. One set of pumps for well and force duty.

Almost certainly the last stationary engine made by this historic firm of marine-engine builders,
this drove a set of three-throw lifting and forcing pumps which performed the whole duty, and were
driven through 4 to 1 reduction gearing on the right of the house. The engine bed was 6ft 6in wide by
13ft 6in long, and the flywheel 9ft in diameter. The crankshaft was a single forging, without
couplings or separate pins, and the condenser was placed in the water flow. Plain and simple, they
were interesting in that the connecting rod bearings were different, one end a strap bearing, and the
other end of the marine type. They were MWB engines No 537, and in continual use until the
installation of the Hathorn Davey triple-expansion engine in 1923 which, of 205hp raised 2¼mgd.
The Penn engine was steamed by two Easton and Anderson Cornish boilers which, at 20 years old,
were allowed to increase the working pressure from 100 to 150psi.

36   Metropolitan Water Board, Southfleet Station, Kent. One engine   (1953)
The Lilleshall Co, Oakengates, Shropshire, 1898. 1mgd. 250ft. 115psi. 10, 16 and 26in x 3ft. Slide
valves. abt 60rpm. 80hp. 3-throw pumps 12¾in x 3ft.
This, too, was in continuous use for many years, until scrapped in the 1950s after the fitting of
electrical pumps. The engine was set upon a single bed 9ft wide and 18ft long, and the built-up
crankshaft, and placing the valves of the middle cylinder on the top made it very compact. The
Meyer valve originally fitted to the high-pressure cylinder was later replaced by a plain one after
many years of heavy use. The overhead receivers placed between and above the cylinders was an
unusual feature, and simplified maintenance by placing all the steam joints upon the top — they
were often oily and very difficult to reach when underneath. The crankshaft was built up, with
separate pins. The pumps again were placed in the well, and driven by cast-iron gearing of about
3 to 1 reduction. The two Cornish boilers by Easton, Anderson and Goolden were Kent district
No 60–61.

37   Watford Waterworks, Eastbury Station, Herts. One engine   (1954)
Worthington Simpson, Newark on Trent, Wks No 5016, 1920. 21in x 2ft 0in. 110rpm. 2.045mgd.
300ft. 142php. 2-throw pumps 12½ and 17in x 2ft 11in.

This illustrates the latest development in the steam pump, the whole being about 20ft long, and
altogether Worthington Simpson made over twenty sets driven by uniflow engines. The steam cycle
gave high power and low steam consumption, and the gearing allowed the engine and the pumps to
work at the best speed. The pump and engine beds were bolted together, which maintained
alignment. Most uniflow engines were enclosed with forced lubrication, and this open-type design
was unusual. The condenser was placed in the water flow under the floor, with the air pump driven
by a crank within the casing at the end of the crankshaft. The type of trip cut-off again was unusual
It was a very good example of sound modern design, compact and efficient, and with many years of
life left when it was scrapped on the adoption of electric pumps, in the 1950s.

38   Wisbech Waterworks, Marham Station. Two engines   (1957)
Dodman and Co, Kings Lynn, c1912? 0.5mgd. 180ft. 60psi. 9 and 18in x 1ft 6in. Slide valves.
Condensing. 25hp. abt 30rpm. Double-acting bucket pumps.
Although not a gear-driving engine it is another example of a pumping engine by a concern best
known for other types. It was, too, an example of using local capabilities. There were two identical
engines placed back to back, ie with the pumps together between the cylinders, with each piston tail-
rod driving a pump. The engines were typical Dodman design, sound, plain, and simple, yet with
everything needed for reliable service. Each was self-contained, and set upon cast-iron girders 16ft
long which tied each engine and its pumps solidly together. The longitudinal girders were tied
together by cast-iron cross girders at each end. The high-pressure cylinders were fitted with Meyer
cut-off valves, and the layout was very neat. The piping was arranged below the floor as far as
possible, and the twin suction pipes are shown, followed by the delivery pipe, and then the exhaust
from the low pressure cylinder which passed to the condenser beside the air pump which was driven
from the disc crank on the end of the crankshaft. They were standby to the Worthington sets (Plate
13) and the cleanliness was typical of this delightful station.

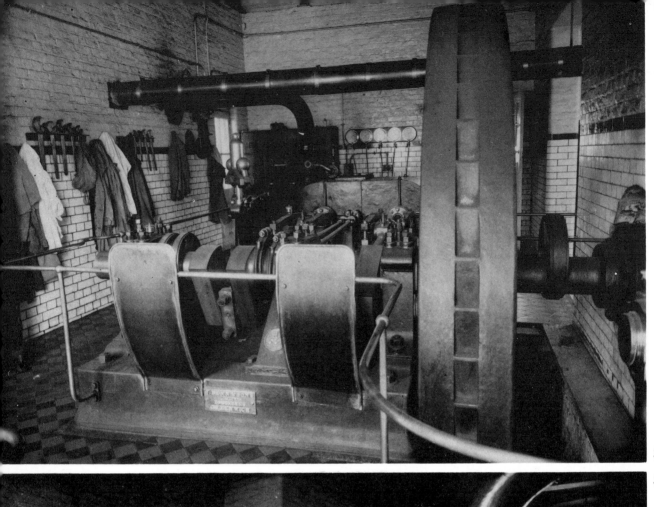

7

8

Vertical Rotative Engines

This was the simplest of all rotative pumping engines, with the cylinder placed on the floor, driving upwards to an overhead crankshaft, and with the pumps below the engine room floor driven directly from the piston tail-rods. It gave a neat engine room with all the piping below the floor, with the engine parts accessible, yet one which, as my examples show, led itself to attractive variations around a simple basic concept, and one which was almost the earliest (with the Cartwright engine of 1797) to break away from the traditional beam-engine form.

39   Weston-super-Mare Waterworks. One engine   (1932)
James Simpson and Co, Pimlico, 1871. 0.75mgd. 150ft. 40psi. 12 and 21in x 2ft 3in. 40rpm. Slide valves. 2 pumps 9¼in x 2ft 3in.

Water supply to the town was started in 1853 by a private concern. This was with a pumping engine known as *Old Bill*, which, since it was said to be heard a distance away, probably had a cast-iron gear drive, and the need for the Simpson engine indicates the growth of the town and water demand. The two sides of the engine were identical except for the cylinders, with plain 'A' frames with the crosshead guides bolted on, and the crankshaft bearings bolts at the top held in place by cotters through the bolts and frame. The six-arm flywheel 8ft in diameter was in two parts held by internal spigots in the rim, and strong cross struts of cruciform design held the 'A' frames rigidly together, to give room for the flywheel, and stay the top. The mechanism was readily accessible, with the marine-type connecting-rod ends widely used by Simpsons at the time. The crankshaft centre was about 10ft from the floor.

40   Grimsby Waterworks. One engine   (1935)
James Watt and Co? c1884. 0.75mgd. 200ft head. 100psi. 10 and 20in x 18in. Slide valves. 80rpm. Three-throw pump, sizes unknown.

Although of the same basic type, this set contrasted in most ways with Plate 39. Regrettably no positive data was available (the above is however near from outside checks) as the plant had long been standby to electrical pumps. It may have been jet condensing, with the air pump driven from the end of the crankshaft and Meyer cut-off valves were fitted to both cylinders. The double-throw crankshaft was a single forging, and the Laird-type crossheads worked around single-bar guides. The flywheel was in one piece, and, possibly only intended for short periods of running, the gearwheel teeth driving the pumps were all of cast-iron. The full steam jacketting of the cylinders and covers, and placing the high-pressure slide valves inside the frame and its fast running speed made it economical. It was neat and well finished as Watt's designs usually were.

41   Liverpool Waterworks, Aubrey St Station. One engine   (1938)
Hathorn Davey and Co, Sun Foundry Leeds, 1896. 2.88mgd. 90ft. 180hp. 140psi. 16, 24 and 36in x 3ft Corliss valves. abt 20rpm. 3 x 22in plunger pumps.

This was the first triple-expansion rotative engine made by Hathorn Davey. It had long been a standby to electric sets, and differs greatly from my other example. The crankshaft bearings were set upon girders built into the house, and were tied to the bed by forged front columns. It was an economical design with the Corliss valves placed in the heads, reducing the clearance, and giving a simple cylinder casting. The flywheels, although light, gave steady turning and it must have been a quiet and pleasant engine to work with. Twin cranks and rods were used to drive the camshaft for the Craig Corliss valve gear, placed at mid-cylinder level. Pumping to the service reservoir was by a 22in plunger pump driven by the piston tail-rod of each cylinder. It cost £4,015 with two boilers, and gave reliable service for many years. It was the forerunner of a large number of Hathorn Davey's other types of triple-expansion engines which served communities all over the world and were built over some 40 years.

39

Inverted Vertical Rotative Engines

In this type the cylinders were placed at the top and the crankshaft at floor level, with the pumps below and driven by side rods passing beside the crankshaft. Adopted for marine and small stationary service in the 1840s, it was adopted for pumping by Moreland in 1868, to become the most popular type which was built up to the largest sizes.

42   St Helens Waterworks, Kirkby Station. One engine   (1953)
Robert Daglish and Co, St Helens, Wks No 890, 1890. 2.5mgd. 350ft? 100psi. 35 and 60in x 6ft 0in. Piston valves. 20rpm. 2 well pumps 20in x 4ft. 2 force pumps 16in x 6ft.

This was a fine example of combining civic pride with local craftsmanship. The foundry work and decorative finish were sheer artistry, with the coats of arms and mouldings picked out in a fine colour scheme, which had needed little attention since it was installed. The high-pressure cylinder was fitted with one, and the low pressure with two piston valves with attractive guides for the valve rods at the bottom. The eccentric rods were also fine forgings with the rod, forked top-end, and the top half of the eccentric strap in one piece. There were no crosshead guides, and the four circular side rods passed beside the cranks to the force pumps below, with the connecting rod returning upwards to the crankshaft, from the force-pump rams. The wells were at the back of the engines (as was done in other engines for the St Helens Waterworks) with the pump rods driven by links and beams from the top of the pump plunger. They were thus clear of obstruction by the engine, and readily lifted by the crane hook at the back. The cranks were placed at 180 degrees apart. It was 20ft high and the flywheel 16ft in diameter. There could have been little profit in the price of £5,590 with two Daglish boilers.

43   Metropolitan Water Board, Wanstead Station. One engine   (1952)
Yarrow and Co, Poplar, 1904. 2.3mgd. abt 170ft. 140psi. 20, 32 and 53in x 3ft 6in. Corliss valves. 210hp. 12rpm. 4 well pumps, 3 force pumps.

This was made by a concern specialising in lightweight high-speed marine engines. The wells were disappointing, yielding less than desired, and two of the four Glenfield and Kennedy well pumps were later disconnected, but the station was valuable and was used as much as possible. The steam valves were in the cylinder heads and operated by Yarrow's improved trip gear, which was basically the Dobson type. It was a well-built engine with cast-iron supporting columns front and back, and was 23ft high to the top. The Glenfield's well pumps were driven by a two-throw extension crankshaft, with one rod to the first quadrant, and twin ones coupling this to the other, but the nearer set were not in use. The wooden rods were fine examples of carpenters' handwork, finished with the adze, and the pump quadrants, metal strapping, and ends were all fine forging. The engine house and condition in which it was all kept was a tribute to a great authority, which still maintains the highest standards.

44   Bristol Waterworks, Chelvey Station. One engine   (1932)
The Lilleshall Co, Oakengates, Shropshire, Wks No 272, 1923. 4mgd. 330ft. 160psi. 20, 35 and 56in x 3ft 6in. Drop valves. 260hp. 24rpm. 2 well and 3 force pumps.

This was the last steam pumping engine of the open reciprocating type built for the Bristol Waterworks Company, and was fitted with a two-throw well pump crankshaft similar to Plate 43 but placed directly over the well, without quadrants and drive rods, and again with force pumps below the cranks. Although marked No 5, it did in fact follow six beam engines at the station, and was very neat, with no couplings between the cranks, a sloping forged-steel supporting column to each cylinder in front, and a cast-iron column behind. It was a very sound engine which would have given many more years of economical service after it was superseded by electrical pumps early in the 1960s. A large engine, over 20ft high, it was also an economical one, since the drop valves were able to handle steam superheated by 150 degrees, and the steam was also reheated between each cylinder in tubes within the circular chambers seen behind the middle platform. Three boilers 30ft long x 7ft 6in diameter were provided, but it would run on one.

42

44

Inverted Vertical Triple-Expansion Engines (Slide Valves)

The high efficiency required from engines which ran non-stop for months on end was usually secured by the use of Corliss or drop valves operated by highly sophisticated valve gears to reduce the heat and pressure losses within the cycle. It must therefore seem odd that simple slide valves, without these advantages, were sometimes adopted for such a service. The plates illustrate instances where this was done, to give engines which ran quietly, were economical, and needed few repairs. Units of this type were appreciated by sea-going engineers who secured good results with such engines in steam ships.

45   Ilkeston and Heanor Waterworks, Whatstandwell Station. Three engines   (1954)
Tangyes Ltd, Birmingham, 1903. Duty unknown. 140psi. 16, 24 and 36in x 3ft 0in. 150hp. 20rpm. 3 ram pumps.

These were standby units in the 1950s, when together with a 1932 Bellis triple-expansion engine and gear driven turbine pumps they were replaced by electric sets. They differed from the following two examples in that the three cylinders were separate, and without an entablature across the tops of the columns. It was really three separate engines as the deep bed was also in three sections. The horizontal air pump and condenser above the floor were unusual features, as also was placing the middle access platform at the back with stairways between each crank. There was a variable cut-off valve to the high-pressure cylinder only. Neat, reasonably economical, and inexpensive, with three units they were very well suited for the service required of them, and certainly had given little trouble, as of course products of a concern with the wide engineering expertise of the makers would suggest.

46   Metropolitan Water Board, Ferry Lane Station, One engine   (1951)
Harvey and Co, Hayle, Works No 805, 1894. 6½mgd. 3 force pumps only. 140psi. 21½, 33 and 54in x 3ft 3in. 230hp. 18rpm. Piston and slide valves. 3 ram pumps.

Another example made by a concern which also made marine engines, this was more heavily built than Plate 45. It differed in several ways, having only a single forged front column to each cylinder, 6½in diameter by 11ft long, no couplings between the cranks, and with the crankshaft at a lower level. The valves differed in that the high pressure one was of the piston type with an internal cut-off valve, but with single flat valves for the intermediate, and low pressures. The maker's oval nameplate can be seen on the cylinders above the middle platform, and the lead sheet covering the bed where the engineman might stand when working around the engine, was an instance of the care taken with waterworks engines. This was fitted with an ejector condenser at one time, and possibly altered later. The flywheel was 15ft in diameter, and the engine 20ft high, steam being supplied by three Lancashire boilers made by Harveys.

47   Brighton Waterworks, Falmer Station. Two engines   (1952)
Fleming and Ferguson, Paisley, 1904. 2¼mgd. 140psi. 18½, 32 and 51in x 3ft 6in. Slide valves. 18rpm. 180hp each. Well and force pumps.

These engines named *Stafford* and *Lowther* after council members were open, attractive sets with features of the marine engines which the makers built. They were not massive, and the twin columns supporting the front of the cylinders were nearly 9ft long. The crankshaft was in three sections, each with its own two bearings so that there were six for the engine, and four for the well pumps beyond. An unusual feature was that all of the valves were of the flat or box type, whereas it was usual to fit a piston valve to the high-pressure cylinder to reduce the friction. Another interesting point was that each cylinder was fitted with variable cut-off by hand-wheel adjustment. The condenser was placed in the water flow, and the air pump was driven by the disc crank at the right-hand end. The wooden casing over the cylinder lagging and the flywheel arms was fine craftsmanship and most attractive. Steam was supplied by three boilers by Penmans of Glasgow.

West Gloucestershire Water Company

The West Glos Water Company supplied water to, and drew its supplies from, a large area within the county, and each of the stations noted pumped from abandoned mines within a distance of some seven miles. By 1920 however the demand was still growing, which led to their drawing supplies from another site some twelve miles away where the power was diesel engines. The steam units illustrate different solutions to a similar problem, that of drawing from mines and delivering through long mains.

48 Frampton Cotterell Station, near Chipping Sodbury. One engine   (1935)
Easton, Anderson and Goolden, Erith, 1900. abt 2mgd. 250ft. 150psi. 16, 27 and 44in x 3ft. Slide valves. 25rpm. 150hp. 2 well and 3 ram force-pumps.

Eastons regularly fitted slide valves to their engines, even for triple-expansion units and this was one of five similar engines which comprised almost the last large engines built at the Erith Works (the other four were for Antwerp). The high-pressure cylinder was fitted with a piston valve with an internal cut-off, with Meyer valves for the outer two cylinders. Steam jackets were fitted to the cylinder barrels and covers, and generally the design was typical of the simplification Easton adopted. The circular motif in the sides of the platforms and stairs had long been an Easton's feature. The engine was 12ft high from the floor, and the air pump was driven by a cast-iron disc crank on the end of the crankshaft. The well pumps were worked by gearwheels of 3ft and 6ft diameter, and a 6ft diameter cast-iron disc crank, to the 19ft long wooden sweep rods (which were 16in square).

49 California Station, Longwell Green. One engine   (1933)
The Pulsometer Engineering Co, Easton and Anderson Branch, 1914. 0.48mgd. 340ft. 100psi. Engine: Robey and Co, No 33529. 12 and 20in x 2ft 2in. 120rpm. 120hp.

This, the Water Company's last steam plant, was probably the last to bear Easton's name, which was cast upon one cylinder cover, almost eighty years after they started in 1836. Because Easton were then largely consultants and designers, the engine was made by Robey, of a standard type which was readily obtained, with drop inlet and sliding exhaust-valves, bored guides, and cast-iron disc cranks. The pumps were driven by belts: the well set by a pulley on the crankshaft to a countershaft and gearing down to 15rpm; the force pump, a Pulsometer turbine type, being driven from the 8ft diameter flywheel by a 14in wide belt, at 1,000rpm. The four well-pumps were driven by wooden sweep rods 16ft 6in long, from the two-throw cranks to twin quadrants.

50 Cowhorn Hill Station, Oldland. Two engines   (1933)
Summers and Scott? Gloucester? 1910. 1mgd. 250–300ft. 80psi. 15½, 21 and 21in x 2ft. Slide valves. 20rpm. 75hp. 1 well and 3 force pumps.

These were very unusual engines in that (a) they were three-cylinder compounds, and (b) the slide valves, including a Meyer cut-off for the high-pressure cylinder, were all worked by a linkage from the cross-heads of the other cylinders, so that there was really no valve gear. There were no eccentrics and by placing the valves upon the top the engine was very compact, little more than the width across the three cylinders, and the two engines fitted into what was probably the old colliery winding-engine house. The ram force-pumps were driven by piston tail-rods, and the well pumps off single cranks on the ends of the crankshafts, with the air and feed pumps driven from a further crank from that. The three force pumps gave the steady water flow necessary with the long mains.

51 This shows the layout of these remarkable engines, with the tail rod pumps, and the cylinders close together. The crank driving the single well-pump can be seen beyond the flywheel, leading to the pumps in the old colliery shaft outside. The close arrangement of the bearings and cranks, with no space taken for the usual valve drive eccentrics is evident. The small disc crank on the outer end of the crankshaft drove the condenser air-pump. The valve gear was remarkable in that there were four valves with different timings, driven by links from the crossheads of other cylinders to give correct timing. There were three cylinders with the high pressure in the middle, exhausting to the outer two low-pressure cylinders. The crosshead of the right-hand cylinder drove the valve of the left-hand one through the central shaft, and the other two crosshead links drove the main and cut-off valves of the central cylinder, and the single valve of the right-hand cylinder through tubular bushes running upon the centre shaft.

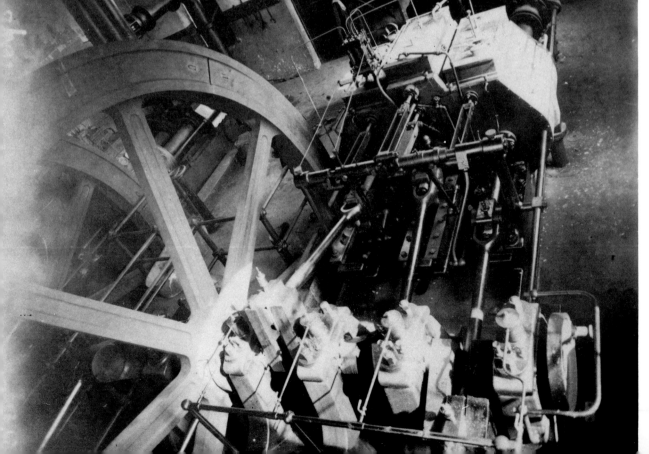

Unusual Waterworks Engine Designs

Waterworks engines usually followed the simplest design that would give reliability with economy of fuel, and so the layout tended to follow set trends. Occasionally however particular service demands, the desire to give all designs a trial, or the need for reasonable economy with low cost, led to departures from the usual forms as these three examples indicate.

52   Nottingham Waterworks, Boughton Station. Two engines   (1957)
Ashton, Frost and Co, Blackburn, 1907. 3.0mgd. 342ft. 150psi + 100°F. 25, 41 and 65in x 4ft 0in. Corliss valves. 250hp? 16rpm. 2 well and 2 force pumps.

These engines maintained the Nottingham tradition for fine engines, with a design combining a horizontal HP with vertical IP and LP cylinders. The horizontal HP was coupled to the centre crank and drove the well pumps behind it by a piston tail-rod, to wooden sweep rods and quadrants similar to Plates 43 and 48. The vertical cylinders each drove a force pump below the cranks by side rods. The short stroke reduced the length of the engine, but it was over 60ft long, and the width was reduced by placing the layshafts for the eccentrics in the front, so that the width was little more than that of the crankshaft and its bearings. They were nearly 28ft high. The four steel front columns and cast-iron back ones with extended feet, made the whole rigid. It was a credit to all concerned, the designers and builders, those who ran them, and the Nottingham Authority. The starting platform with two levers in a frame added to the attraction of the design.

53   Metropolitan Water Board, Hammersmith station. No 13, The Leavitt Engine   (1948)
James Simpson and Co, Pimlico, 1901. 3mgd. 260ft. 100psi. 18 and 36in x 7ft 6in. Corliss valves. 195hp. 14rpm. 2 ram pumps, 15¼in x 7ft 6in.

This type, so called from the American engineer, E. D. Leavitt, who adopted it in the latter part of the nineteenth century, embraced vertical cylinders driving through a bell crank to a horizontal connecting rod and crankshaft. No 13 had the pumps placed below the engine room floor, and driven by side rods from the crossheads. The steam valves were in the cylinder heads to reduce clearance, and the cylinders were fully steam jacketted which with the long piston stroke reduced the steam consumption. The Leavitt was a compact type, requiring less space than a vertical or horizontal engine of the same power. With the parts mounted upon a deep bed, and the pump load taken directly from the piston rods, it was a rigid design, but despite this No 13, even with counterbalance in the flywheel rim, did rock at the top at full load. Two Oldham Ironworks Co boilers with superheaters for 100°F, gave economy which kept the engine in constant use for over thirty years.

54   Stockport Waterworks, Wilmslow station. Three engines   (1960)
Frank Pearn and Co, Manchester, c1900. 0.36mgd to 200ft. Force pumps only. 2x9 and 15in x 1ft 0in. Slide valves. 35–40hp.

The 'Banjo' pump, sometimes so called from the shape of the casting which coupled the steam and pump ends together, and in which the connecting rod vibrated, was widely used for auxiliary service. When as here it was fitted with compound steam-cylinders, worked with early cut-off, and two were coupled together with a heavy flywheel, the fuel consumption was greatly reduced so that it was acceptable for a small waterworks such as Wilmslow. The far engine, a twin tandem-compound set over 17ft high, with an 8ft diameter flywheel, was a very powerful unit of the type. The nearer ones (there were two), were cross compound with cylinders about 12 and 21in x 1ft 3in stroke, again with cranks at 90°, and were nearly 10ft high. The Pearn engines were the forcing units handling the water pumped by the well set (Plate 33), and were good examples of a basically simple and widely used design developed for economical waterworks use, to give economy with reliability.

53

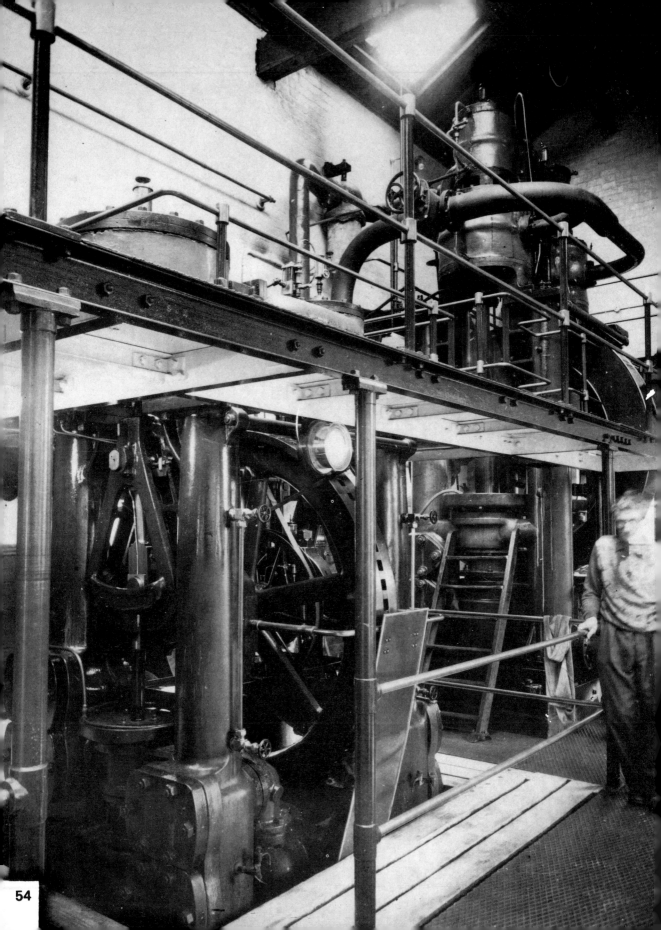

Three more unusual waterworks examples, including a table engine, and an oscillating engine, types almost unknown in this service.

55 Eastbourne Waterworks. One engine (1952)
R. Moreland & Co, London, 1895. abt 1mgd. 240ft. 100psi. 17 and 31in x 2ft 6in. Slide valves. 23rpm.

The inverted vertical-tandem engine was not a common type, especially in waterworks pumping service, and this was the last of four or five such engines supplied to Eastbourne. This type of direct rotative engine with the pump below, driven by two or four side rods from the crosshead was introduced by R. Moreland and D. Thompson in 1868. The design was sound, with the cylinders supported upon rigid 'A' frames upon a deep bed, to which the pump system was coupled by a cast-iron framework below, with side rods straddling the crankshaft. The steam cylinders were coupled close together, with the HP below, and a box gland between them, each having Meyer cut-off valves. It was a large engine, over 14ft high, and was fitted with the bucket and plunger pump favoured by D. Thompson, which together with the heavy flywheels, 10ft in diameter, gave steady running even with the single crank. The plant was on the shore below the cliffs, with two Moreland boilers of 1896, but was disused in the 1950s.

56 Gravesend Waterworks, Lieth Road Station. No 2 engine (1937)
Maker and history unknown. 21in x 3ft 0in. Slide valve. abt 25rpm? 3-throw well pumps.

This was the only table engine I found in waterworks pumping service, and was also the largest example of the type I met. The cylinder was fitted with a single slide valve and a cam driven auxiliary drop valve with cams for three cut-off points. The pumps (similar to Plate 25) were driven through 3 to 1 reduction gearing. The table supporting the cylinder was 6ft high by 5ft 9in square, the side connecting rods were 10ft long, and the fluted columns supporting the table were 4ft 6in high by 7in diameter. In every way it was a most attractive engine which had given good service, and was steady running as there were no stays to support it from the building. An unusual feature of the 16ft diameter flywheel was that the hub was bored, with a single key, whereas most table engine flywheels were of large bore fitted with stakes, upon flats on the shaft.
    The No 1 Engine was a Woolf compound beam-type by Easton, Amos and Anderson, of 1869. The pumps had long been electrically driven, however, and little was known of the steam plant.

57 Metropolitan Water Board, Hammersmith (1936)
Harvey & Co, Hayle, 1870. 5½ x 10in. Slide valve, non condensing.

The site workshop was an important part of the early pumping stations, where the first superintendent was often the engineer who erected the plant. He was a skilled craftsman and so able to do repairs with the large machine tools often installed with the pumping machinery. This engine probably drove the workshop plant supplied with the later Bull engine (see Plate 8), at Campden Hill Pumping station in 1870. It was typical of Harvey's sound design, and although small, the slide valve was driven by a sliding sector to give it the best action, in contrast to the simpler but less efficient valve gear of many small oscillating engines. This engine is now preserved in the Science Museum, London.

56

Sewage Disposal

This was the second of the great problems affecting the daily life of the town dwellers in the rapid growth of the nineteenth century. The insanitary conditions of the Middle Ages are well known, yet were tolerable with the lower population. The great growth of the towns in the Victorian era, when expanding industry in towns rapidly demanded a local populace of operatives, produced a problem of unprecedented magnitude. Epidemics of cholera and other diseases occurred with a mortality and frequency that demanded attention. This involved massive sewer construction to allow collection at a central point and for pumping to treatment plants away from the towns. It was a challenge which the Victorians met as they did every other, with cautious consideration and, once committed, with immense energy, yet with careful financial consideration.

Recycling waste is no new concept. The scientist Liebig calculated that the sewage of London in the 1850s contained annually some 46,500 tons of dry matter, sufficient to manure 1,000,000 acres of land, and even then worth some £350,000 per year. In 1843 on this point, a contractor agreed to pay £22,000 per year for the sewage of Paris, then to be deposited five miles farther away from the city.

Sewage disposal did not automatically mean that all was pumped — great consideration was given to gravity movement. The vast London Main Drainage Scheme of the 1860s comprised three distinct main drains, a high, middle and low, on the north side of the Thames, which met at Abbey Mills, and three on the south, meeting at Deptford pumping stations. Such was the skill of the planning, however, that the high and middle level sewers were high enough to drain by gravity to the final discharge points at Barking on the north side of the Thames, and Crossness on the south, the main pumping being that of the low-level matter into the upper. Another station was needed to drain the extensive low-level area of West London, which at Chelsea raised the sewage some eighteen feet into the low-level drain.

Inland towns without large rivers had to provide treatment plants, and at Leicester, and Burton-on-Trent, four large compound beam pumping-engines were installed, since the areas were too low for more than local gravity clearance. Another method for low lying areas was the pneumatic lift in which sewage was collected at several points to be raised by compressed air in ejectors to drains to the disposal point. Typical of the responsible attitudes of the times was the Victorian precept of making power for sewage lifts by incinerating refuse to make steam to compress air for the ejectors. This was done at Eastbourne.

The nature of sewage, and the limited capacity of the mains, required continuous pumping, and the load varied greatly, especially in wet weather. Great flexibility and economy was thus essential, and the pumping stations had two to four engines available most of the time. The pumps were usually of the single-acting ram type with hanging valves; the use of centrifugal pumps for sewage removal developed at the end of the nineteenth century. The need was to keep town drains clear and pump the waste matter to treatment plants a few miles away, against a head of 30 to 50 feet, and as my examples show, the solutions to this basic need were indeed varied in type and make.

With sewage and surface water clearance dependent upon pumping, heavy rainfall had to be moved quickly to avoid flooding in low areas, and special storm-water units were installed to deal with it. Centrifugal pumps were very suitable for this, and in east London steam-driven storm pumps were in use at West Ham, beside the Lilleshall beam engines, together with gas-engine driven pumps in other stations where there was no steam plant. Split into multiple units, pumping could cope with the heaviest storms, so that the great floods of the nineteenth century with water several feet deep in large areas of the towns are now almost unknown.

Beam Engines

Examples illustrating differing approaches to the problem of driving large low-lift pumps placed either side of the beam centre.

58   London Main Drainage, Western Station, Chelsea. Four engines   (1931)
James Watt and Co, Birmingham, 1873. 9mgd. 18ft head. 35psi. 37in x 8ft. Drop valves. 10rpm. 90hp. 2 ram pumps, 66in x 4ft.

Western was the last of the four main drainage stations to enter service and was the only one to remain unaltered throughout its period of steam service. The engines at Abbey Mills (eight) and Deptford (four) were made compound, and Crossness (four) triple expansion about the turn of the century. The Western engines show the high standards of design and workmanship which were applied in the scheme, and they needed little other than running repairs in 60 years of service. The pumps were placed either side of the beam centre, and driven by massive cruciform rods which, together with the connecting rods, platform supports and railings, were artistry in cast-iron. The great rocking beams overhead however, were built up from plate and angle iron, and the slight creak of the rivets was the only sound when they were at work. The highly decorative casing of the centre supporting structure may have partly been of timber, as was the casing for the cylinder — otherwise everything was high class engineering in cast- or wrought-iron.

60   Tottenham Borough, Markfield Road Station. One engine   (1964)
Wood Bros, Sowerby Bridge, 1886. 3.7mgd. 20ft head. 80psi. Cylinders abt 21in x 4ft and 45in x 6ft. Piston valves. 16rpm. 100hp. 2 ram pumps, 26in x 4ft 3in.

Nineteen engine builders tendered for these engines, and the contract was awarded to Wood Bros at £2,450. The design with eight fluted columns was very attractive, and all of the decorative work was in cast-iron. One pump was driven directly from the beam, and the other by a piston tail-rod below the high-pressure cylinder, which left little balancing to be done by the 28 ton flywheel. The piston valves were of Wood's rotating variable cut-off type for the HP, and simple rotation for the LP cylinders, with manual cut-off adjustment. The beam again was built up from plates, but the engine was entirely independent of the house. The valve chests on the left of the cylinders were attractively ornate, and again the cylinder lagging timbers had no brass retaining strips. When added capacity was needed in 1905, three Worthington engines each with cylinders of 8, 12 and 20in diameter were installed, with 27in pumps, running at 38 double strokes per minute.

61   Burton-on-Trent Council, Clay Mills Station. Four engines   (1936)
Gimson and Co, Leicester, 1885. 3mgd. 110ft head. 80psi. 24in x 6ft and 38in x 8ft. Drop valves. 150hp. 10½rpm. 2 ram pumps, 21in x 6ft.

These engines, called A, B, C, D, were arranged in pairs in houses on either side of the boiler plant. They were generally plain, yet neat, and certainly gave good service, with one major overhaul in seventy years of working. The beams were built up of iron plates and one of the pumps was driven from a piston tail rod, and the other near to the connecting rod by stiff rods and parallel motion. They ran very well, with little noise, even from the iron teeth of the wheels for the governor and valve-gear drive. Although mostly constructing small engines and intricate machines, Gimson's ventures into the large beam-engine field did them great credit. There were latterly five Thompson boilers, two steaming two engines, with mechanical stokers, and there were numerous small engines for ancillary duties on the plant.

62   Borough of West Ham, Abbey Mills. Two engines   (1936)
The Lilleshall Co, Oakengates, 1900. 20ft head. 120psi. 30in x 4ft 9¾in and 48in x 8ft. Slide valves. 8–10rpm. 2 ram pumps, 63in x 4ft 9¾in.

These were powerful engines pumping to the main London discharge tanks, and were later aided by centrifugal type storm-water pumps. All were steamed by nine boilers. The plant had to be able to cope with any demand at once, and the centrifugal pumps were in the adjoining house. Again, one of the ram pumps was driven from the beam and the other from the piston tail-rod. Other than the simple motif upon the entablature cross girder there was nothing spent upon decoration, but the whole was substantial. A very unusual feature was that the flywheels were built in three pieces, each with three arms, while the Meyer cut-off valves were driven from the parallel-motion linkage, with a single eccentric on the crankshaft driving both main valves. A good feature for engines which had often to stand, ready for work for long periods, was the steam jackets on the covers as well as the sides of the cylinders.

59   London Main Drainage, Deptford Station. Four engines   (1931)
Stothert, Slaughter and Co, Bristol, 1864–6. 2,500cfm. 18ft head. 35psi. As built: 48in x 9ft 0in. Drop valves? 2 ram pumps per engine. 84in x 4½ft. Altered by Hathorn Davey in 1905 to Woolf compounds. 22in x 6ft 8in and 38in x 9ft.

Bristol has not figured in my record so far, but it did share the most important city drainage scheme of the nineteenth century, when Stotherts' tender for four engines at £22,000 was accepted for Deptford. Actually due to alterations in the specifications, the final price was agreed at £23,400. They were good engines which were heavily used, and the photograph shows that there was a degree of individuality in the finish and decorative style, when it is compared with Plate 58 of Western Station in the same scheme. The railings at Deptford were especially attractive. The beams were single-web castings 32ft long by 5ft deep, with top flanges 20in wide and the drop valves were placed at the side of the cylinders, as far as can be traced, whereas Watt and Co placed theirs at the front or back of the cylinders. Ten Cornish boilers were installed originally, but these were replaced when, at the turn of the century, all but the Western plant were remodelled, with new cylinders and high-pressure boilers. The new cylinders at Deptford were made by Hathorn Davey and Co, with Corliss valves. Otherwise the engines appear to have remained unaltered and in use until replaced by oil-engined pumps in the late 1920s; the beam engines were used into the 1930s. Although Bristol was not a centre for large engine building, it was a tribute to Stotherts that they could cast the 32ft beams and make the 26ft diameter flywheels.

60

Sewage Pumping — Other Types

63 Hull Corporation Main Drainage Station. Three engines (1957)
James Watt & Co, 1883. abt 3mgd? 20ft head? 100psi. 18 and 30in x 4ft 0in. Piston valves. 150hp. 18rpm. 2 double-acting bucket pumps, abt 3ft 6in x 4ft.

These engines were named *Hollitt, Willows* and *Massey*, after local councillors, and drained an ever-increasing area of Hull for over 70 years. Typical of Watt's practice, they were plain but sound engines, with the pumps driven by the usual side rods and with the connecting rods driving upwards from the lower crossheads. The pumps, made by Thornton & Crebbin of Bradford, were connected to the engine bed by cast-iron columns, 8in diameter. The HP cylinder was fitted with a Rookes-type internal piston valve with variable cut-off, and a plain one for the LP. The flywheels, 12ft 6in diameter, were very heavy in keeping with the rest, and the engines stood over 12ft above the floor with the pump well 30ft below it.

64 Poole Corporation, Main Drainage Station. Two engines (1958)
Goddard, Massey & Warner, Nottingham, 1902. abt 2mgd. 100psi. 15 and 2 x 16in x 2ft. Slide valves. 30rpm max. 50hp. Ram pumps, 12in x 2ft.

These were interesting as they were non-condensing, the small size and very variable load not justifying the fitting of condensers. Three cranks allowed them to run very slowly (down to 5rpm). The HP cylinder at the left was fitted with a Meyer type variable cut-off, and the other two with plain slide-valves. The engine beds were 20ft long, the flywheels 10ft diameter and the cast-iron balanced disc cranks were 2ft 6in diameter. The exhaust steam was used to heat the feed water for the Cornish boilers. In every way it was a good plant for the service, simple yet able with little waste to meet the very variable demand: 8–10rpm being suitable for dry weather, while wet weather needed up to 30rpm. An interesting feature was that there were only three bearings for the three cranks, suggesting that each cylinder largely absorbed its own load.

65 Wellingborough Corporation. Irthlingborough Road station. Two engines (1965)
Tangyes, Birmingham, 1895. Duty unknown. 100psi. 10 and 16in x 2ft. Slide valves. 60hp. 90rpm. Double-acting bucket pumps, 9in x 2ft.

These illustrate a completely different approach to the last plant, in having a tandem-compound condensing engine running at a fast speed, driving two pumps through 4 to 1 reduction gearing. It was very compact and by nesting the engines and pumps together the overall size was greatly reduced, yet it was quite accessible, and low in cost by using almost standard range units. The double-acting pumps with cranks at 90 degrees gave a steady fluid flow, and with the wooden teeth in the large gear wheel must have run very sweetly. The apple wood teeth were 7in wide and held in place by fitting wooden wedge fillers between the tails. The flywheels were 8ft diameter, and the engine and condeser was 20ft long over the flywheel, and the pumps were about 10ft long. A by-pass valve in the exhaust pipe allowed the engine to be worked non-condensing at starting or for other needs. Differing in almost every respect from the previous example, they were ideal for a growing community, reasonably low in cost and operation.

66 Corporation of Eastbourne, Central Compressor Station. Two engines
Hughes and Lancaster, Ruabon, 1892. 1470 cubic feet air per minute to 20psi.

10in x 2ft. Slide valves. 75hp. Compressors. 15in x 2ft, up to 80rpm?

Low lying seaside towns presented a special problem with sewage removal, since a central sewage collecting point with gravity delivery might involve very deep sewers. The Shones air-lift system overcame this by lifting the sewage a short distance by the direct application of low pressure air, so that movement was affected by gravity and re-lifting. It was simple and effective, and was excellent at a time when small centrifugal pumps with simple drives were not available. It was made the more valuable at Eastbourne by siting the compressor station close to the refuse destructors, so that the steam produced by incineration was used to compress the air, which in turn kept the town and its drainage system healthy by sewage clearance. There were two of these very compact units each with twin steam and compressing cylinders at opposite sides of the 'A' frame. The flywheels were 7ft in diameter, with the crankshafts 12ft above the floor.

4

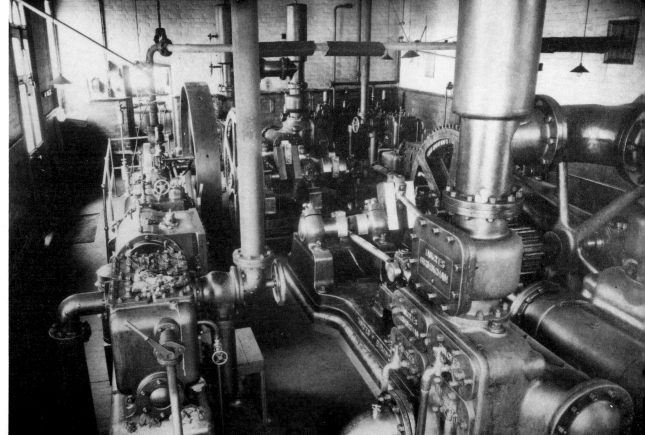

5

Land Drainage

The effective control of water to the soil is essential for good farming, which only survives in the East by irrigation, ie by adding water to the land. In England the reverse is the case, with large areas of fertile ground too low for natural drainage which, until the coming of steam power, could not be adequately farmed owing to flooding. The first steam drainage units were the traditional large beam engines and flat-bladed scoop wheels similar to those long driven by windmills. Two or three of these (similar to Plates 67–69) still survive, and a single beam engine and scoop such as that preserved at Stretham in Cambridgeshire, would maintain good farming over an area of several thousands of acres, whereas a dozen or so windmills would only do so inadequately owing to the irregular action of the winds. As an example of the situation prior to, and showing the impact of, steam power, Thomas Neale writing of Cambridgeshire in 1748, noted the ruinous state of the Manea area, and that there were in all about 250 windmills in the whole middle area of some 160,000 acres. Whittlesea alone had fifty, and nearby Doddington almost as many. In one 6 mile stretch he saw forty windmills. Land drainage brought its own problems since the land sank as it was drained, and scoop wheels could not then reach the water, and this led to the proliferation of such wheels. Thus the Littleport and Downham area of some 26,000 acres contained seven windpumps in 1770, but by 1820 there were eighty there, and they were costly to maintain and build. Replaced by two large beam engines of 1819 and 1830 with scoop wheels of 43ft diameter and a further small plant, one of the 43ft wheels had to be increased to 50ft diameter later. It is probable that this area contained the greatest concentration of windmills, some of which lasted for nearly a century.

The efficiency of the scoop wheel was much increased by controlling the intake and discharge by curved breasts which, admitting only the amount of water that the flat scoop blades could move effectively, thus prevented them from wallowing in dead water. The heavy engine and scoop wheels required massive foundations and were costly to install, and as the land sinks as it is drained, this required the diameter of the scoop wheel to be increased to reach the water at the lower level of the land. Despite these disadvantages, however, the scoop wheels were retained for many years as they were simple, gave good service, and could be maintained by the men who ran them, and as Plates 69–70 show, they were sometimes adopted even for small areas.

A major development in land drainage occurred with the introduction of the centrifugal pump from 1850, and the two leading concerns in this were Eastons and Gwynnes. This type, revolving at 80 to 150rpm, was much smaller and lighter than the scoop wheels which ran at 4 to 8rpm, and being less costly allowed authorities of small areas to improve the drainage. Although their pumps were generally similar, Gwynnes and Eastons differed widely in their arrangement of the pumps and driving methods.

Although they were used for the great Dutch drainage scheme of the 1840s, displacement pumps of the bucket or plunger type were little used for the purpose in England where it was probably limited to Waldersea, Norfolk. There an area of about 5,050 acres was drained by a 70hp Cornish beam engine made by Harveys of Hayle in 1831. The steam cylinder of 40in and the pump of 6ft in diameter were of 8ft stroke. This was replaced in 1883 by an Easton engine and pump (see Plate 68)

The Archimedian screw revolving within a circular casing was an early device which also played a part in land utilization in the nineteenth century. The last large installations were probably those of the 1880s at Atfeh and Katabieh in Egypt, which raised 8,000 tons of water per minute for land irrigation. In 1881 Easton and Anderson made the Katabieh set comprising ten screws, each raising 8,000 tons per hour 12ft high. Water elevating screws were also used for land drainage in Europe and one is preserved in Denmark. This, once driven by a windmill, was later worked by an old horizontal slide-valve engine, and finally by a highly efficient drop-valve engine working with superheated steam. Other water-raising screws were also used in Holland, and the engines were usually horizontal. A few such screws were used in England, driven by windmills.

Beam Engines and Scoop Wheels

Three examples of differing applications of the same basic layout, these were the first reliable land drainage units, massive traditional designs which needed only fuel and attendance to keep large areas of land fertile. Developed from the type long used with windmills, the scoop wheel failed when the land level sank with drainage, and it had then either to be increased in diameter by fitting scoops of greater length, or be replaced by one of greater diameter. Few scoop wheels were built once the utility of the centrifugal pumps was established in the 1850s.

67   The Soss Drainage Station, Misterton, Notts. Two engines   (1937)
Booth and Co, Park Ironworks, Sheffield, 1839. Drained 10,000 acres. Called *Ada*. 10–20psi. 33in x 7ft 0in. Slide valves. 85hp. 18rpm. Scoop wheel 34ft diameter x 4ft 6in wide.

Nothing was known of the other engine which had long been replaced (see Plate 77) but *Ada* was a useful standby. An attractive engine, the foundry work of the railings and governor standard were neat, as were the fluted trunks connecting the top and bottom valve chests. The beam had double webs, was 20ft long and the flywheel 18ft diameter. The engine was unaltered, even retaining the soft packing for the backs of the 'D' shaped steam valves.

68   Chear Fen, Cambridgeshire, Cottenham and Rampton Drainage Board. One engine   (1937)
J. Clark and Co, Sunderland, 1842. Drained about 5,000 acres. 30in x 5ft. Slide valves. 18rpm. abt 50hp. 10–20psi. Scoop wheel 28ft diameter x 1ft 8in wide.

This engine shows several detail differences to Plate 67 as it had wooden engine railings, a single web beam, and a wrought-iron connecting rod which was forged with a neat centre boss. The 20ft flywheel was also much lighter, with flat instead of 'T' section arms. It drove through cast-iron spur gearing with a 5 to 1 reduction, and was standby to an oil engine in 1937.

69   Sutton Marsh Drainage, Halvergate Station, Norfolk   (1938)
Fletcher and Co, Derby, 1861? Area drained unknown. Disused 1930. 50psi. 9in x 2ft 0in. Slide valve. 35rpm? abt 10hp. Scoop wheel 18ft diameter x 6in wide.

Most of the drainage engines were pumping units only, and so were often idle for most of the time. At Halvergate, however, some at least of the idle time was utilized, as it also drove a single set of flour-grinding stones by a 3ft 6in diameter belt pulley on the crankshaft. The little plant was as simple as possible, the engine, which was 6ft 6in high, being non-condensing, with an exhaust steam feed water heater jacket around the exhaust pipe. The Cornish boiler was made by J. Walley, Derby.

Small Scoop Wheels

Two examples of this type built to serve small areas long after the centrifugal pump had proved its capabilities. These show the simple steam plants used to drive them, which although they are of familiar types, were rarely adopted to drive scoop wheels.

70   Morris's Engine. The Sixteen Foot Bank, Stonea, Cambridgeshire.
Make unknown. abt 9in x 1ft 0in. Slide valve. 130rpm. 14hp.

This combination of an overtype engine and a scoop wheel was probably unique in English land drainage; it was also modern, possibly made after the turn of the twentieth century. The engine was a standard non-condensing overtype design, and with the boiler feed-water heated by the exhaust steam. The scoop wheel which was made by J. Varlow of Benwick, was 27ft outside diameter, with paddles 5ft long and 9in wide, and was made largely of rolled steel. It was driven by two-stage cast-iron gearing with a total reduction of 24 to 1 giving a scoop wheel speed of about 5½rpm.

71 and 72   Middleton Towers Estate, near Kings Lynn   (1938)
J. C. Baker, Kings Lynn, 1877. 6½ x 9½in. Slide valve, non-condensing. 10hp. 120rpm. 50psi.

71   This little plant served to drain a low level area within the Ramsden estate, transferring the water to the main drains in periods of very wet weather. Again it was as simple as possible with the non-condensing engine driving by single reduction gearing to the scoop-wheel shaft. Steam was supplied by a vertical boiler 3ft 6in diameter x 8ft 0in high. There was no feed-water heater, the exhaust steam pipe passing straight to the roof. The feed-water pump was originally driven by the engine and fitted on the bed plate, but was replaced by an injector in later years.

72   The scoop wheel was 16ft tip diameter, and fitted with forty paddles each 3ft 3in long x 5¼in wide. The gear reduction was 1 revolution of the wheel to 15 of the engine.
The scoop-wheel framing consisted of a series of identical castings bolted together along the arms, with slots in the rim for the paddles.

Scoop Wheels and Centrifugal Pumps

The scoop wheel comprised a massive wheel with numerous flat paddles or scoops attached to the rim. Rigidity was essential since the paddles had to fit tightly to the sides of the stone breast in which they worked to reduce leakage, and a first-class wheel was as close as half an inch each side, so the rim was supported by massive arms, sometimes with cross-braces between the arms. The paddles were supported by rigid 'starts' of square timber set in sockets cast in the rim. The effort of lifting the water stressed the starts and paddles severely, so they were stayed to each other by circular rings of flat iron, which spread the loading over the whole ring of starts. For best effect, the paddles were placed closely together, and were set at an angle of 25 to 40 degrees to the radius, so that they met the water quietly with little loss by splashing. They worked best with a tip speed of about 8ft per second. The largest English wheels were at Hundred Foot Bank, at first 30ft, and altered to 40ft, which weighed 54 tons, and was later replaced by one 50ft diameter weighing 75 tons, which raised 213 tons of water per minute at 3rpm.

73 Deeping Fen Drainage, Pode Hole Station. Two engines   (1936)
563 tons per minute.

The larger wheel was 28ft diameter, later increased to 31ft, with forty paddles 6ft 6in long x 5ft wide. It raised 300 tons per minute. The print shows how the diameter was increased by extension bolted to the starts. The mechanism of the outlet rising delivery shuttle is at the bottom right-hand corner.

74 Sutton Marsh Drainage, Halvergate Station, Norfolk.   (1937)
Duty unknown. 18ft 6in tip diameter. Paddles 6ft long x 6in wide.

An attractive small country wheel, with a cast-iron frame 10ft diameter. This was driven by the engine shown in Plate 69. The engine was small and ran fast, with double-stage reduction gearing in the timber house on the right.

That water might be raised by centrifugal force was known in the seventeenth century, but probably not applied until the Massachusetts centrifigual pump in 1820, and then at a New York dock in 1830. Revolving at high speeds, ie 120–150rpm, they were much smaller and lighter than scoop wheels, and the actual pump rotor (Plate 75) probably weighed 5 per cent of the scoop wheels they replaced. The pump was completely enclosed in a cast-iron casing.

75 Rotor: Easton, Amos and Sons, pump, 1864

This pump drained the 3,000 acres of Curry Moor in Somerset for over 90 years, and could raise 90 tons per minute 3–5ft high. It was a magnificent piece of bronze casting 4ft 6in diameter and nearly 17in deep between the outer flanges. It was a single casting, demanding the highest skills of the moulders and foundrymen to produce sound castings of the complex hydrodynamic curves of the Appold pump, especially with the rapid transition from the thick central boss to the thin vanes and outer flanges, which were ⅜in thickness. Driven at about 120rpm by an engine similar to Plate 79, with cylinders 20in x 2ft, it used 4 tons of coal per day, and in one very wet period ran for 380 hours non-stop.

76 Curry Moor Drainage, Somerset. 3,800 acres   (1950)

A view of the pump well of the Amos 1858 patent combined unit. A rare photograph, this shows the vertical shaft down to the submerged pump rotor, and the protective sleeve around the shaft carried up to discharge level. At the right are the wooden automatic flap doors to the discharge channel. The well chamber was fitted with the integral cast-iron lining which can be seen to pass around the valve chest of one cylinder, to protect it with high water levels in the drain. Passing down inside this case was the pipe carrying steam to the valve chest, controlled by the cock on the chest at the top. The section of the well lining was separate, and removable for cylinder maintenance.

73

74

75

76

Gwynnes Pumps

There were two Gwynne concerns building low-lift centrifugal pumps in London in the latter half of the nineteenth century. They were John and Henry Gwynne at Hammersmith, and Gwynne and Co at Essex Street, Strand, and later at Brooke Street Works, Holborn. Each provided a general engineering service, but were quite separate, in fact Gwynne and Co's advertisement in 1868 stated 'Please address in full to prevent mistakes'. Each used the simplest engines of the horizontal or inverted vertical types, with the pump directly connected to the engine crankshaft. For irrigation, the engines were usually horizontal compound condensing, such as those for the Ferrara scheme in Italy of the 1880s, which made 200 square miles of swamp fertile. For dock service, however, the engines were horizontal or inverted vertical, simple or compound, condensing or non-condensing, to best meet service requirements. The two concerns joined forces at the end of the century, to become Gwynnes Ltd (J. and H. Gwynne and Gwynne and Co, United) at Hammersmith, until after about a century they became Allen-Gwynnes Pumps of Lincoln. J. and H. Gwynne built some of the largest plants for irrigation, drainage and dock pumping, and an article of 1895 describing a pump installed at Lynden, Holland, noted that 'in 25 years they had built pumps able to move 42,550 million gallons of water per day or about 140 times the mean flow of the River Thames'. Later as Gwynnes Ltd they continued to manufacture very large pumps for dock and drainage service, including the 1930s three units for the Middle Level drainage scheme in the Fens, which with 102in diameter delivery pipes, could together pump 2,000,000 gallons per minute to a height of 11ft.

77   The Soss Drainage, Misterton, Notts. *Kate*   (1937)
J. and H. Gwynne, Hammersmith, c1890? 30in pump. 75 tons water per minute maximum. Abt 12 and 22in x 1ft 6in. Slide valves, condensing. 135hp. 75rpm.

There were two quite separate plants at The Soss, originally with beam engine and scoop wheel, and this plant replaced one similar to '*Ada*' (Plate 67). Only occupying a part of the old engine house, *Kate* was able to drain the whole area except in very wet weather, and did so into the 1950s. Nothing was known of the engine which was replaced by *Kate*.

Easton, Amos and Anderson — A Brief Record

Eastons were associated with the water industry for most of the nineteenth century, beginning in 1820 when Josiah Easton secured the English manufacturing rights for Montgolfier's hydraulic ram, and which he greatly improved. He joined his first partner Mr Leahy at Grove Works, Southwark in 1827 and continued there when Leahy left in 1829.

C. E. Amos became a partner in 1836, and the works developed to make a wide range of machinery for waterworks and factories. The adoption of Appold's centrifugal pump from 1850 brought business in land drainage machinery which was greatly increased by the adoption of C. E. Amos's patent design of a combined engine, well casing and pump unit. This led to the institution of many small drainage commissions, since it was inexpensive, and easy to install. They became Easton, Amos and Sons in 1860 and William Anderson joined in 1864. His first task was to plan a new works at Erith, equipped with the latest machinery, now an urgent need as the business had outgrown the Grove Works. Anderson's influence was soon felt, with new engine designs (Plates 80, 81, 82) and the new works were able to undertake large work, among it the machinery for *HMS Rover*. This was a large contract, worth nearly £80,000, for a horizontal three-cylinder engine and ten boilers. The boiler busisness grew until by 1900 they had supplied boilers to over 300 customers, many with several per order, together with a wide range of engines and machinery. Easton and Amos retired in 1866, but the business continued to grow. Design became ever more plain and machine-finished, with straight lines replacing the curves of the earlier ones. The manufacturer of gun mounts in the 1880s was probably due to the armaments knowledge of William Anderson, who retired from the concern upon his appointment as head of the Government explosives department in 1889. They remained Easton and Anderson until 1895, when Mr Goolden joined, and as Easton Anderson and Goolden they developed an extensive business in electrical plant, including generating equipment for eight or nine public authorities, as well as electrical and hydraulic lifts, electric motors and pumps. They continued to make waterworks machinery, and their last large contracts were for waterworks triple-expansion engines, including four for Antwerp, about 1900. A group of engineers then bought the works to make, as Easton and Co, the Schmidt high-superheat steam engine, but it was probably too advanced, as although they secured large orders, the business did not develop, and by 1905 they were again Easton and Anderson largely as consultants and designers.

The illness of the leading partner in 1908 and the loss of a senior staff member led to the sale of the business to the Pulsometer Co of Reading, to remain as 'the Easton and Anderson branch' for a number of years. Plate 78–82 illustrate the progressive simplification of Easton's smaller land drainage engines, and in the 1890s a few vertical-shaft pumps had the balanced horizontal-engine coupled directly to the top of the pump shaft.

Easton, Amos and Anderson's Drainage Engines

78   West Butterwick Drainage, Lincs, South Common Area   (1936)
Easton, Amos and Co, London, c1855. Drained 700 acres. 30–40psi. 14½in x 2ft. Slide valves, condensing.

Eastons probably made engines of this grasshopper type for factory driving by 1840, and this was an early application of the type for driving the centrifugal pump of 1852. It was characteristic of the artistic designs of the period, with attractively curved castings and forgings. The pump was outside, driven by bevel teeth on the flywheel face to a pinion and horizontal shaft, and then by another pair of bevels to the vertical pump shaft. Twin-cylinder overbeam grasshopper engines were rare.

79   The Somerset Rivers Drainage Board, Westonzoyland Station.   (1936)
Easton, Amos and Sons, 1861. Drained 2,000 acres. 60–70psi. 20in x 2ft. Slide valves, condensing.

This was the design covered by the patent No 2802 of 1858, combining the engine and pump unit with a cast-iron well lining, and the pump placed at the bottom of the vertical shaft. This made the whole an integral structure, fully fitted and ready for simple erection on the site. The parallel-motion guides, with condenser pumps driven from the main levers were early features of the design, and construction was largely the work of the foundry, blacksmith, turner and fitter.

80   Somerset Rivers Drainage Board, Aller Moor Station, Burrowbridge   (1935)
Easton, Amos and Anderson, 1869. Drained 1,926 acres. 60psi. 13½in x 2ft. Slide valves. 60hp? 50rpm.

This mechanically simply design needing better machining capacity reflects the influence of William Anderson, and the new works at Erith with its modern machinery. This type was less costly to make, with the engine non-condensing, gaining power from the higher steam pressure available from the improved boilers. It was another step in Easton's process of simplification of design, with the best use of the metal, and machine finishing largely replacing handwork. The removable section of the cast-iron well casing, which gave access to the cylinder and valve chest, is clearly visible on the right hand engine. All 1858 patent type units had this for each cylinder.

81   Somerset Rivers Drainage Board, Stanmoor Station.   (1937)
Easton, Amos and Sons, 1864. Drained 790 acres. 40psi. 11in x 2ft. Slide valves. 40hp? 50rpm. Condensing.

This was probably the only engine which Eastons built to this design, which required several alterations to the frame and well-top structure of the 1858 patent shown in Plate 79. Simple slidebars replace the parallel motion, with the cylinders inclined to drive to the single crank. The boiler-feed pump was driven from the expansion-valve eccentric which drove both expansion valves, with the single air pump driven from the other end of the crankshaft. These were features which would have reduced the cost had the design been widely adopted.

82   Waldersea, Norfolk. One engine   (1937)
Easton and Anderson, Erith and London, 1883. Drained 5,050 acres. 80psi. 18in x 2ft 6in. Slide valve. 45rpm, pump 135rpm.

Illustrating the continuing simplification of Easton's designs, this did have a condenser, but the whole was mounted upon a continuous bed, which was inclined upwards to meet the crankshaft centre line at 6ft above the engine room floor. The vertical pump shaft was driven by a 90 tooth mortise-bevel flywheel, to a 30 tooth cast-iron pinion on the vertical pump shaft. The flywheel was 10ft diameter and the cast-iron disc crank, simple connecting rod, and crosshead, and the ram-type condenser air-pump gave economy with the reliability and limited attention so necessary in land drainage.

78

Electricity Supply

The virtues of electricity were recognised almost as soon as it was established (a) that a current was generated by spinning a coil in a magnetic field, and (b) that the process was reversible, so that the power produced could be transmitted to drive a similar unit as a motor. The value of electric lighting was recognised equally quickly, and in the late 1880s the use of a fairly simple arc lamp to illuminate a large area was employed for many colliery pitheads, theatres, workshops, and for street lighting. The first engines used were often portables and undertypes, driving to countershafts and then to the dynamo by belts to secure the high rotational speed required by the generators.

The demand for electricity grew rapidly so that bulk supply from central, rather than individual private stations, was necessary and justified. After a century of effort the invention of the Parsons turbine and generator in 1884 gave the direct rotating engine that had been a dream for so long. The turbine developed rapidly over the next twenty years, as did engines and generators. Almost at the same time as the Parsons turbine was introduced, a reciprocating engine, the Willans, first patented in 1874, was developed in 1884, to couple straight to generators at 350 to 500rpm, and with its valves in the centre of the piston, worked economically. It was the most popular type of generating engine until, early in the twentieth century, demand outstripped the capacity of these engines, and turbines led the immense development of the electrical industry. It was not the end of the private generating station, however, especially for institutions, and all types of hospitals. These, often isolated, had no local bulk supply available, and they could use the low grade heat of the exhaust steam for the services of hot water and heating. The Willans engine was widely used for this, and a number remained in use, as main or standby sets, into the 1950s.

The enclosed, forced lubricated engine introduced by Belliss and Morcom under Paine's patent of 1890 also entered the field, again with high economy and effectiveness, particularly where the heat of the exhaust steam could be used. Nearly 10,000 were built, up to 2,500hp per unit and many remain as valuable standby sets today.

The electric tramways and railway systems provided cheap and rapid transport to more spacious suburbs. This was another instance where the steam engine provided more than the simple mechanics of life, by giving better living conditions for town dwellers, as well as giving easier movements within the towns. Development was rapid, and the urgent need for generating equipment came at a time when the need for such equipment was heavy in other directions. American manufacturers, with a large home demand for electric traction had developed well tried and reliable systems and plant. Large open-type engines with flywheel generators upon the crankshaft gave good results in this service, and were often retained as long as such transport was used. The urgent need occasionally led to the importation of American plant, as in the case of the Mersey Railway, but many English engines and generators gave very good service. Often the enclosed forced-lubricated engines were adopted, as with the splended Summer Lane generating station of the Birmingham tramway system with some 20,000hp of Bellis and Morcom engines, and generally British manufacturers ably met the sudden demands. Public electricity supply was at first provided by a variety of engine types, and the coming of the Willans engine was a great help to the engineers when larger and more economical types were needed to cope with the very rapid increase in demand. The Willans type was used in every supply service, and my examples show how little the external appearance altered; there were, however, numerous design changes to meet the need for economy and larger units, including variable cut-off and triple expansion. The demand for large units for public transport, able to handle extreme and rapid variations in demand was often met by massive open-type engines. My photographs do scant justice to the generating engines, but at least illustrate the trends at the turn of the century.

83   The Herrison Mental Hospital, Charminster, Dorset. Four engines   (1950)
Willans and Robinson, 1895(3) and 1904(1). 12 and 30kW rating. 3 engines 10ins x 5in; 1 engine 12in
x 6in. abt 450rpm.

The hospital, situated in a delightful but isolated site, was an early user of carbon filament lamps,
and with no public supply, met its needs with these engines for nearly 60 years. Simple expansion,
they were able to work on low-pressure steam, and exhausting to calorifiers, greatly assisted the
supply of hot water and space heating. Well cared for, the main change was from copper gauze to
carbon dynamo brushes, in the 1920s.

84   Avonbank Power Station, Bristol. Four engines   (1958)
Triple expansion and compound, two and three cranks. 1902.

Willans engines provided the first power for the Bristol tramways in the 1880s and for public supply
from the Temple Back Station of 1892. That station was extended to the maximum to meet the
demands, and by 1902 with nineteen engines the station was filled, and a new site was found at
Avonbank, again with Willans engines at first, and Plate 84 shows their appearance. Two years
later, however, turbines had to be adopted to meet the immense growth of load, but the Willans units
remained for house service as long as the station was used.

85   Shore Road Generating Station, Mersey Tunnel Railway. Three engines   (1948)
British Westinghouse Co, Manchester, 1,200kW. 1,650hp. 650volts AC or DC. 30 and 60in x 4ft. Drop
and Corliss valves. 85rpm. 160psi.

The large power stations of the turn of the twentieth century were splendid places which often
showed the reciprocating steam engine at its best, and Shore Road was typical. The engines were
27ft high and with cylinders at 21ft centres, the overall width was 27ft. The generator stators were
over 20ft in diameter, and with the engines spaced at 25ft centres down the house they really were
impressive. The engines were probably made in the USA; certainly it was an American design with
the high-pressure cylinder drop-valves in pockets at the sides, and the low-pressure Corliss valves
placed in the cylinder heads. They supplied the power when the railway was converted from steam
to electric traction, and were used with Parsons turbines until power was taken from the Grid in the
1950s. The boiler house with nine Stirling boilers was also very fine with mechanical stokers, and the
chimney 250ft high was a local landmark, even where chimneys were greatly in evidence. The
engines were probably the last large open-type power station sets in the UK, although the
magnificent Manhattan engines of New York subways survived some years after Shore Road was
closed.

Gas Supply

The earliest gas supply was from small local stations serving a limited area, so that mains were not extensive, and the natural pressure built up within the retorts was sufficient to deliver the gas to the consumer. The convenience and cleanliness of gas for heating and lighting soon led to its greater use, and it was then necessary to service much larger areas from a single station. The extended mains produced so much resistance to flow that it then became necessary to increase the pressure at the supplying station. For the essentially local systems the pressure was, however, quite small, possibly a few ounces per square inch, and, to give finer control this was expressed as 'inches of water'. Usually, a single-stage eccentric sliding-vane exhauster or booster of the Beale type sufficed for a small works, pulling the gas through the system from the retorts and delivering it to the storage gas holders. But soon the large works required this to be in two stages, one to pull the gas from the coking retorts through the various stages of purification, and with another similar low-pressure booster to deliver the gas to the storage holders under a pressure sufficient to deliver it to the distant mains and the consumer. This basic pattern sufficed for over half a century with a town station supplying its own area and with quite small country towns having their own plants.

The competition of electricity and the need to reduce costs in the twentieth century, particularly after 1920, made many of these small units too costly to run, leading to their closure. The consumer demand had then to be met from the supply systems of the large towns, and the extending mains from these created a resistance to the gas flow that required greater pressures to deliver gas to the small local areas. This was beyond the capacity of the simple eccentric-vane Beales type boosters, so piston-type blowing units, usually driven by high-speed enclosed engines came into use, to provide gas pressures of three to five psi, which would feed through mains many miles long. The engines for this were usually non-condensing, and frequently twin-cylinder rather than compound. Turbine-driven fan type boosters were also adopted in the last era of the coal gas works. By far the greater number of gas works steam engines were thus of the simplest open horizontal non-condensing slide-valve type, running at 20rpm, or more except for heavy demand periods.

In view of the serious danger from loss of pressure, when pilot lights would also fail, ample standby booster units were provided, usually with two at work at once so that in the event of a breakdown of one, the other could be speeded up to maintain the load until a standby engine could be started. The speed and capacity was determined by a gas pressure operated governor placed immediately at the outlet from the booster. Gas works were valuable to the community for providing considerable employment, much of it semi or unskilled labour; an example was the Windsor Street Gas Works in Birmingham which, situated beside railway and canal, was in 1897 making ten and a half million cubic feet of gas per day. The plant had fully mechanised coal and coke handling, necessary indeed since over 190,000 tons of coal were carbonised per year. Despite this, the works employed some 750 men including a gas meter servicing department, and an experimental coal carbonising test plant and laboratory. The coal was carbonised in 750 retorts, and the gas was handled by five Beale-type exhausters. Each of these pumped about 150,000 cubic feet per hour maximum, and were driven by two Waller horizontal condensing-engines of 16 inches bore by 20 inches stroke, a Robey horizontal engine, 14 inches by 20 inches, and one Allen tandem-compound condensing engine of 16 and 25 inches bore by 18 inches stroke to drive the other two. Ten Babcock and Wilcox water-tube boilers and three Lancashire boilers, fitted with furnaces to use waste coke dust as fuel, supplied all of the steam. The use of the then waste coke dust, or breeze, was general in gas works for steam raising, and almost all auxiliaries were steam driven at low cost. Very small country works often used gas engines for such drives.

As a rule each exhauster or booster was driven by its own steam engine, as my examples show, but at the Old Kent Road works of the South Metropolitan Co they were driven by belts from a line shaft. The earliest types were single-cylinder beam engines, followed by compound beam-engines in 1866–70, and later by the Bryan Donkin horizontal engines Due to their high speed, gas engines often drove boosters by belts and line shafts. Most of the small plant was driven by separate single-cylinder slide-valve engines often in the open air, as were the many small pumps for liquor and tar transfer. The pumps were often of the Evans 'banjo' type.

Gas Works Exhauster and Booster Drives

The photographs illustrate the variety of engine types which were used in the Victorian era when older and newer designs were made side by side, and sometimes an old design of engine in stock could be delivered quickly or cheaply for a small works. Certainly Messrs Waller offered many such designs for service in small works. One of their table engines was used in Canterbury gasworks for many years, and they offered beam, oscillating as well as other types for small works into the 1880s.

86   The South Metropolitan Gas Works, Old Kent Road, London   (1948)
Thomas Middleton, London, 1866. Duty unknown, several units. 60psi. abt 9 and 24in x 3ft. 20hp. 16–18rpm.

The first boosters here were driven by single-cylinder beam engines, and when they were replaced by compounds in 1866, the fuel consumption was reduced by 50 per cent. The Middleton engine was a unique annular compound, and was an excellent example of the high grade London engineering of the nineteenth century. The high-pressure cylinder was inside the low-pressure cylinder, with the three piston rods attached to a single crosshead. The very neat cylinder lagging casing with its wooden retaining bands was unusual. There was only one slide valve. It drove by a 10inch belt.

87   The South Metropolitan Gas Works, Old Kent Road, London. *Gamma* engine   (1948)
Bryan, Donkin and Co, Bermondsey, c1878. Duty unknown. Two engines. 60spi. abt 12 and 20in x 2ft 3in. 30hp at 79rpm rated horsepower.

These followed the beam engines, with two exhausters per engine, driven by belts through fast and loose pulleys. They were of the Farey type, with the low-pressure piston rod passing backward to a crosshead, and side rods to the front one. It was very compact and avoided an internal gland between the high and low pressure cylinders. Again condensing, it also had a variable cut-off for the high-pressure cylinder, giving high economy at variable loads. The condenser air pump was driven from a pin on the connecting rod near to the crankpin. These were standby to other engines of the 1920s.

88   Winchester Gasworks   (1936)
Easton and Anderson Ltd, London and Erith, 1892. Duty unknown, preserved 1936. abt 19in x 1ft 6in and 1ft crank stroke. Meyer slide valve. abt 50rpm.

This grasshopper beam engine also drove by a belt to the exhausters, and with a large cylinder gave good power on low-pressure steam as it, too, was condensing. It indicates the simple and plain characteristic of Easton's later engines, with a double-web beam that allowed the use of single crosshead and connecting rod top-end bearings. The main valve was driven through a tubular shaft, with the expansion valve arbor inside it, and the eccentric rods were circular. It was a charming little engine that had worked hard, and was (in 1936) preserved.

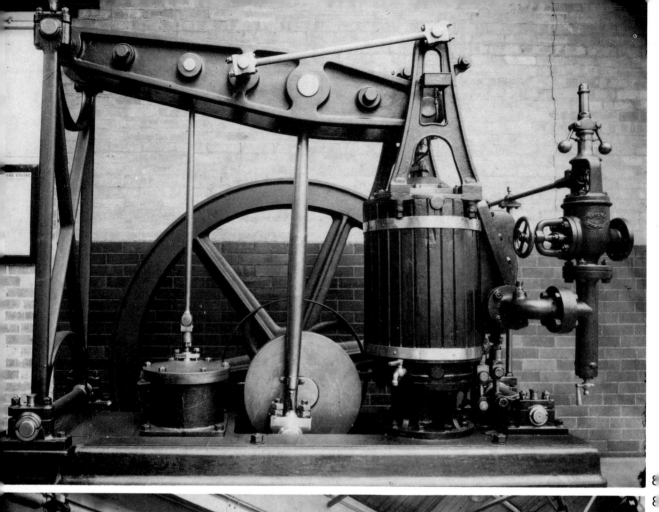

These photographs show the later designs of engines and gas handling units which followed the previous examples of the early period. Plate 90 is a design which, usually with a single Beales exhauster, driven directly from the crankshaft, was met in almost every gasworks. Between them, the photographs illustrate the types that served so long, the later developments of this (Plate 90) and in Plates 89 and 91 the advanced types needed to service the altering load requirements as the industry was concentrating into larger central production units serving ever growing areas.

89   Bristol Gas Co, Stapleton Road Works. Two engines   (1950s)
George Waller and Co, Stroud, c1900s? 60,000 cubic ft per hour. 5psi. Wks No 2384–5. 12in x 1ft 2in. Slide valves. 10–50rpm. Steam 60–80psi.

This photograph illustrates the changing trends in the industry and the machinery needed to service them. The two single-cylinder Waller engines represent late developments of the traditional Beales rotating-vane booster of the 1840s, efficient at all speeds from 20 to 80rpm. The engines were substantial with solid framing and massive parts and bearing areas, which with continuous lubrication would run for months without stopping, and with little attention. The Bryan Donkin engine on the right was a highly developed type with twin high-pressure steam cylinders above the piston-type gas compressors, which were mounted upon the crankcase. Fully enclosed and forced lubricated this would run continuously, and with ability to provide high pressure (up to 25psi), served when outlying works were absorbed and long mains were needed to feed gas into country districts. This unit shows the value of higher speed as at 250rpm it handled four times the volume that the open ones did, at five times the pressure increase.

90   Bath Gas Works, North Engine House   (1950s)
Bryan, Donkin and Co, Chesterfield, c1928? Various engines. 16in x 1ft 6in. Slide valves. Twin boosters of 150,000 cubic feet per hour each. 20hp. 15–20rpm.

This late example of the traditional type shows how little the exterior had altered, but it was different internally with several modifications which increased the capacity as well as the period between repairs, by modifications in the sliding vanes and rotor and casing. The twin boosters increased the capacity per unit, and the 300,000 cubic feet per hour was as large as it was desirable to go with a single unit of the type. There were three similar units of various dates at the works, together with later types. They were fully controlled by the gas pressure through the pressure governor seen beside the engine cylinder.

91   Swindon Gas Works. Various engines   (1950s)
Holmes Roots-type boosters driven by Ashworth and Parker, and Sisson engines.

The Sisson engine on the right dated from 1939, and the Ashworth and Parker was works No 1686 of 1943. They indicate how gas companies were continuously modernising their plants. The Roots-type blower comprised a pair of figure-of-eight shaped rotors which revolved in phase through gearing, the Ashworth revolving at 313rpm and the smaller Sisson at 530. The ring oiler bearings for the rotor shaft bearings ensured continuous running with little attention. The Ashworth, with rotors 3ft 3in long, was very powerful.

The Smaller Ships

The little ships, able to move in difficult waters and handled by simple men of great skills, gave yeoman service for generations. In fact for many years the Clyde 'puffers' as they were called, were the mainstay of the small towns on the Scottish West Coast.

The equally valuable service of the steam tug needs no stressing, nor that of the Admiralty V.I.C. ships usually called 'Vics', which provided shore to ship bulk handling facilities. They were all virtually sea-going errand boys, and in general the power plant was similar — a compound surface-condensing inverted vertical slide-valve engine, supplied with steam by a single small Scotch return-tube boiler. The main variation was that some of the smaller ones had large vertical boilers, with a plain centre flue and water tubes across the firebox, or else Cochran cross-fire tube boilers. There was more variety in the sand trade than the others, and sometimes boats, still sound but outgrown in their own trade, were adopted for sand supply. Cooper's *Alpha* was a case in point.

92   SS *Tredegar*, Messrs Osborne and Wallis, Coal Factors, Bristol   (1948)
G. K. Stothert and Co, Bristol, 1892. Hull 94ft x 14½ft x 8ft deep. 20in x 1ft 6in. Slide valve. 35rhp. 140rpm. 1 two-furnace Scotch boiler. 50psi.

This vessel ran on the short Cardiff to Bristol coal carrying run where ability to meet the tide times was more important than high economy. The single-cylinder engine which was short, saving space and weight, was therefore well suited for the duty which the ship performed for some 60 years. The engine was fitted with a counterbalanced flywheel 3ft in diameter, and the engine did not stop on the centres. The foundry work was exceptional, with the condenser cooling water and exhaust steam, passages cast in as part of the framing and bed. The pumps — air, circulating, feed and bilge — were all driven by levers from the crosshead. Everything was as simple as possible, and the boiler, seen beyond the engine, indicates how short the engine compartment was.

93   TSS *Bulldog*, Port of Bristol Authority, general service vessel. Two engines   (1950)
E. Finch and Co Ltd, Chepstow, 1884. Hull 88ft x 20ft x 8½ft deep. 12 and 24in x 1ft 2in. Slide valves. abt 300hp. 120rpm. Two Scotch boilers. 100psi.

Bulldog was fitted as a crane and grab dredger; she was also a salvage vessel with pumping equipment, as well as a useful tug, and was in service for some seventy years. She was twin screw, and very handy, and the engines were interesting as they were single-crank tandems. Cross-steam connections were fitted, which allowed either boiler to steam either or both engines, as well as the deck crane and auxiliaries. The engines were 10ft high and very sturdy, again, with the condenser pumps driven from the cross-heads, and with 3ft diameter flywheels. The condensers, however, in contrast to Plate 92, were separate, vertical and placed in the corners of the engine room. Also, in contrast to Plate 92, only a minimum of the steam passages were moulded into the framing.

94 and 95   TTS *Alpha*. Cooper and Co, Salford. Two engines   (1948)
Hull built by T. A. Walker, Sudbrook. 145ft x 30½ft x 10¾ft deep. Engines: Fleming and Ferguson, Paisley, 1890. 9, 12, 16 and 24in x 1ft 9½in. abt 400hp. 120rpm. 160psi.

Fleming and Ferguson supplied twenty-three engines of this size and type to Walkers for twenty-one ships (two were twin screw). *Alpha* and her sister *Beta* were built as dredger tenders for the Manchester Ship Canal Co in 1890 and were sold to Coopers in 1938, when the ships were lengthened, and a new boiler was fitted to each. The photographs indicate how compact the design was: 8ft 6in high and a little over 5ft square, and about 8ft 6in centres across the ship. The hardwood casing over the cylinder lagging was fitted without the usual brass bands, and the cylinder blocks were in two castings with a flanged centre joint. The valves and valve gearing were intriguing, as each four cylinders was served by two piston valves driven by two sets of link motion, for which there was only two eccentrics. These were on the crankshaft, and drove upwards to work the centre valve for the IP2 and LP cylinders. The front valve for the HP and IP1 cylinders was driven by horizontal rods from the eccentric straps, to operate the front link-motion by the rocker cranks on the bed. It was most ingenious, and was possible as the engine cranks were at 90 degrees. All of the piston and valve rod glands (six per engine) were fitted with soft packing, and the condensers were a part of the engine frame, with the pumps driven by levers from the centre of the forward connecting rod. The ships were scrapped about 1956.

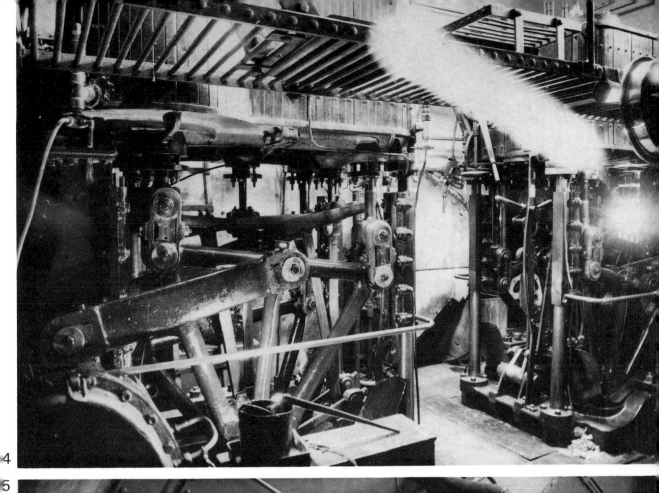

4

5

Leisure Ships

The steam engine also served to make life brighter by its contributions to leisure. Again the railway and fairgrounds are well known, but the machinery of holidays upon the water is less so, hence my inclusion in this review. The splendid paddle steamers, whose excursions to the sea gave diversion and changes of scenery to millions of townsfolk over much of a century, are well known. The passenger steamers which provided a similar service upon the rivers, although less well known, equally gave quiet and carefree days afloat. Much less is known, however, of the engines which drove the vessels, and my photographs of these will be a reminder of the warm oil smell, and almost total absence of noice and vibration that went with steam days afloat.

In the early days the engines of the passenger ships to the seaside, as on the Thames, Clyde and Avon rivers, were varied, with occasional side lever, grasshopper, oscillating cylinder, and on the Clyde steeple engines. These disappeared as the boats that carried them aged, and only the diagonal engine and a few oscillating engines were in passenger service for the seaside runs by the 1950s. A special feature of the light draught paddle-ships was the haystack boiler, a free steaming vertical type of large diameter, introduced by Napier. The complex form of the water spaces in these boilers showed the boilermaker at his best, an artist in iron, able to work plates three eighths of an inch thick, into contours that, it would seem, could only exist on the drawing board, and not in a boiler tight enough to work at 50psi. Such boilers were used mostly for the passenger paddle-ships, often with simple expansion and diagonal single engines, up to 57in bore x 6ft stroke. Very large overhead single-beam engines were widely used in the USA.

The evolution of the river and fresh water steamboat engines also embraced a range of types at first. On Windermere, the twin-screw high-pressure non-condensing engines fitted on *Tern* and her sisters, were horizontal, and placed below the boiler barrel, out of the way. On the rivers there were mostly inverted vertical single-cylinder and compound engines made by a number of concerns, such as Bellis and Seeking, Wilson of London, Plenty of Newbury, and doubtless a few by Thornycroft, des Vignes and others that gave long service. The leader in this field, however, was Sisson of Gloucester, founded by a brilliant engineer, William Sisson, in the late 1880s. They made superb engines, which, beautifully finished and a delight to watch, were of outstanding reliability, and long lived, giving up to sixty years service. They reached their peak with the non-condensing triple-expansion engines fitted to Salter's boats on the Thames, Bathurst's boats on the Severn at Tewkesbury, and many others. Although they did build a few engines with Stephenson's link motion, most of their engines were fitted with radial types of valve gear of their own (Patent No 3634 of 1885) with no eccentrics, or the Marshall or Hackworth types, with a single eccentric per cylinder.

Non-condensing operation was a feature of the river boats, where the light weight, lack of condenser pumps, and ability to use ample cylinder lubrication assisted the necessary quiet operation, and reduced any power loss from back pressure. The boilers, usually single furnace Scotch return-tube, coal fired, gave very good service using river water only for boiler feed. Sisson also made boilers of this type, as well as the Sisson's patent vertical and water-tube designs for some twenty-five years, but the later boilers were usually made by Abbot of Newark.

96 PS *Empress*, Messrs Cosens and Co, Weymouth   (1949)
John Penn and Son, Greenwich, 1879. 25psi. 30in x 2ft 9in. Slide valves, 34rpm. 50rhp. 11 knots.

The ship's hull, built by Samuda at Millwall, was 160ft long x 18ft beam x 8ft 4in deep. The engines were originally fitted with a jet condenser using salt water, and the first boiler was oval, 13ft high x 9¾ft wide x 8⅔ft long, with three furnaces. It was found to be too small, and was replaced, and a surface condenser fitted in 1884, alterations which made her a good ship for many years. The photograph shows the engines when she was on the Weymouth passenger tripping service, extremely well kept, and typical of the best Thames engineering tradition. They were almost noiseless when at work, and were a sheer delight to watch. The link-motion reversing gear made her easy to handle by the handwheel, to the left of which, overhead, can be seen the two steam control valves, one for each cylinder. The circular casting between the two cylinders was originally the jet condenser, into which the steam was discharged through the trunnions upon which the cylinders oscillated. It was regrettable that hard usage during the war and the age of the hull led to her being scrapped when some seventy years old, but the engines are preserved at the Southampton Maritime Museum.

97   PS *Britannia*, Messrs P. and A. Campbell Ltd, Bristol   (1950)
Engines by Hutson and Son, Kelvinhaugh, Glasgow. Wks No 209, 1896. 100psi. 37 and 67in x 5ft 6in.
Piston and slide valves. 42rpm. 2,000ihp 304nhp

Britannia, built by McKnight of Glasgow, was 230ft long x 29ft beam x 9½ft deep, and greatly liked,
was equally regretted when scrapped at 60 years old. She was fast, and the engines served unaltered
the whole time, but new boilers were fitted in 1921, 1935 and 1948, the first being of the haystack, and
the later of the Scotch types. The engines were of the usual diagonal form with massive stays
between the cylinders and the crankshaft bearings, and with the valves, driven by Stephenson's link
motion, placed at the sides. The engines were controlled from the starting platform at the bottom
right-hand corner; auxilliary hand reversing was provided by the large hand wheel in the centre, and
orders were transmitted by the engine-room telegraph which was well lighted by the engine room
skylight. Their splendid condition was a tribute to the old style men who maintained them with a
personal pride in cleaning that gave them a sheen equal to the drawing room silver. There were
many who, like the author, saw more of the engines through the windows than they did of the
seascape.

98   PS *Bristol Queen*, Messrs P. and A. Campbell Ltd, Bristol   (1960)
Rankin and Blackmore Ltd, Wks No 517 (1946). 160psi. 27, 42 and 66in x 5ft 6in. 2,700ihp. 45rpm.
Piston and slide valves.

The ship was built by Charles Hill and Son, Bristol, and at 961 tons gross was the largest paddle ship
in the fleet, ie 244ft long x 31.2ft beam x 10½ft deep. The engines again were very much the standard
form, and were fascinating to watch and travel with, the three cranks giving much less of the fore
and aft plunging that was so marked in the single-cylinder engines. The starting platform in the
centre, although it handles three cylinders, only needed one more control than *Britannia*'s two
cylinders. Again fitted with link motion for reversing, this with short rods and long suspension links,
had the least link slip I ever saw. Throughout, the engine was a credit to everyone, the builders,
owners and the engine room staff, and her short life of some twenty years was regrettable. Her one
double-ended Scotch boiler was always oil fired.

99   SS *Henley*, Salter Bros, Oxford   (1955)
W. Sisson and Co Ltd, Gloucester, Wks No 501 (1896). 180psi. 6½, 8½ and 11in x 7in. Piston valves.
350rpm. 80hp.

Salter Brothers' vessels were interesting in that although the engines were mostly triple expansion,
they were non-condensing, the exhaust passing to a silencing tank and overboard, or to the funnel.
The piston valves were at the front of the engine, driven by W. Sisson's patent gear valve, by which
the valves were driven entirely by linkage, ie, there were no eccentrics. When running there was a oil
splasher plate at the front, removed to take the photograph. The general layout was the same in all of
the dozen or more boats in the fleet, with a single-furnace Scotch boiler, coal fired. *Henley* was in the
charge of William Arnold when the photograph was taken, and the care shown is typical of the
veteran river engineer who, as so often, not only lived with but loved his engines. As with PS
Empress firing was really artistry, the boiler never blowing off steam to waste, yet was kept within a
few pounds of the maximum, which was necessary with a non-condensing triple-expansion engine
on a tightly timed river run.

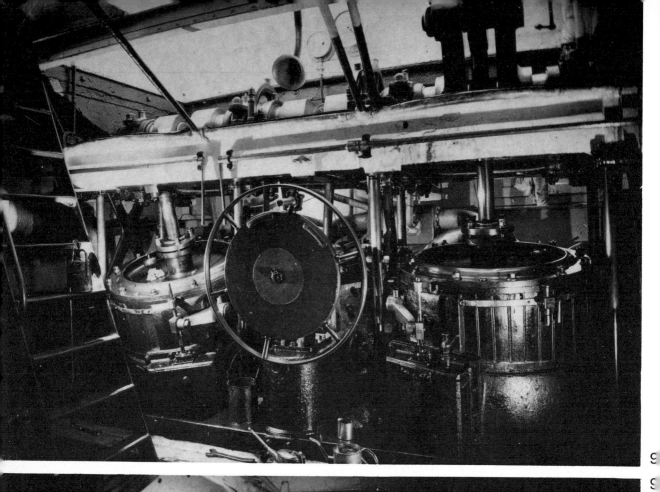

Tunnels, Docks and Floating Bridges

The Severn Tunnel was one of the many great feats which the Victorian engineers performed with simple equipment, and it has always required heavy pumping. Construction started in 1872, but work stopped when a great spring of water was encountered in 1879 and to master this required one of the greatest concentrations of pumping machinery ever used in tunnelling, and heavy pumping has always been necessary. The permanent installation comprised twelve Cornish beam engines and two Bull engines, all made by Harveys of Hayle, and forty-eight boilers were provided to supply steam to these and the fan. The main installation was six engines pumping from the 29ft shaft, but I have chosen two lesser known units. The Bull engines were saved when the steam plant was removed following electrification in the 1960s.

100   Severn Tunnel, The Iron Shaft pumps, Sudbrook, Mon.   (1950)
Two Bull Engines: Harvey and Co, Hayle, 1876. 50in x 10ft. Plunger pumps, 26in diam.

One beam engine: Harvey and Co. 75in x 10ft. Bucket pump 38 then 35in x 10ft.

These two Bull engines were the first permanent pumps to be installed, and were at work in 1878. They were installed at one side of the Iron Shaft (so called from its cast-iron lining) and the 75in beam also pumped from this shaft. This was an 'odd man out' for the tunnel as it was the only one with a cylinder bore exceeding 70in, and also it was the only one of the 'waterworks' type, ie with the beam centre bearings supported upon four ornamental columns instead of a solid masonry wall. This was an advantage in that the house provided the rare panorama of a Cornish beam engine (foreground) and two Bull engines (background) in a single view. Nine single-flue Cornish boilers provided steam at 40psi for the three engines.

101   Severn Tunnel, The Ventilating Fan Engine   (1950)
Walker Bros, Pagefield Ironworks, Wigan, 1923. Abt 15 and 30in bore x 2ft stroke. Up to 90rpm.

The original fan was of Walker's Guibal type, 40ft diameter by 12ft wide, which was driven by a slide-valve engine with a cylinder of 33in bore and 2ft 9in stroke. Installed in 1886 it ran at some 30rpm. The new plant was also made by Walkers, consisting of tandem-compound engine coupled directly to the fan shaft. The engine was a highly efficient drop-valve type fitted with the Doerfel positive valve-gear operated through twin side-shafts. With the cylinders mounted upon a single cast bed, the crankshaft bearings carried in a strong frame, and the two connected by the crosshead guide-trunk, it was an economical engine which operated effectively until it was replaced by twin motors driving to the fan by vec belts.

An important dock service was the provision of water under a high pressure for the operation of dock gates, coal tips, moving bridges, and cranes and capstans.

102   Newport Docks, South Wales. One engine of this make and others   (1955)
Galloway and Co, Manchester, 1914. 1250gpm to 830 psi. Pump rams 7¼in diameter. 28½, 45 and 70in bore x 3ft 0in stroke. Drop valves. 40rpm. 1600hp?

These engines represent the peak of steam-driven hydraulic pumping-plant and worked with an economy almost equalling that of waterworks engines. Installed during a general improvement of the docks, they dealt with a constantly varying load with the positive regularity that ensured the rapid turn round of shipping loading coal by the continuous operation of the wagon tips. The large size and high steam specification for pressure and temperature, and rapid response to loading fully repaid the cost of the plant. The double-ram pumps were driven by twin piston tail-rods, and with accumulator and speed governor control any load was met by one engine on full, one on partial load, and one on standby.

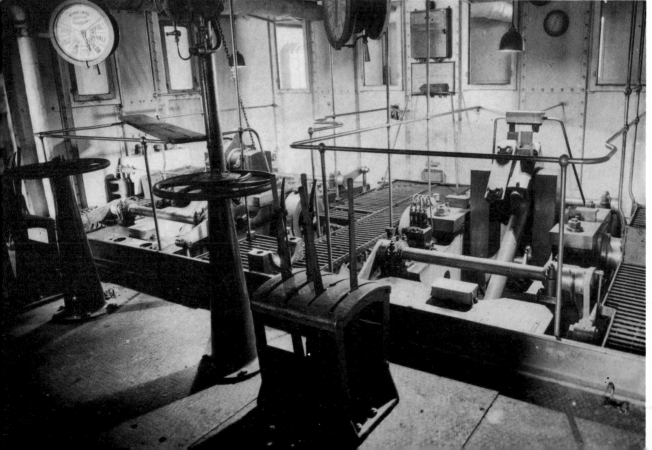

The floating bridge is a neat solution of the problem which arises when a busy road is intercepted by a waterway through which a fairway must be maintained for shipping. They were really floating platforms which traversed the waterway to carry passengers and vehicles from side to side, and were provided with sloping slipways or a floating platform to allow working at any state of the tide. They were of two types: the free vessel as Plate 103, which was moved by paddle wheels or screw propellers driven by the usual marine-type engines, and the chain or cable ferry, which traversed a chain or cable laid across the waterway, the chain falling back into the water as the bridge was wound across it.

103   The Woolwich Free Ferry, River Thames   (1948)
J. S. White and Co, Cowes, 1922 to 1930. Four cylinders 33in x 2ft 9in stroke. Slide valves. Abt 500hp, 45rpm. 30psi.

The ships and engines were similar to those of 1889 which these replaced. The paddles were independent, each driven by a simple double diagonal-engine, in charge of an engineer with orders from the bridge. The original engines were built by John Penn and Sons and fitted with air pumps driven by levers from the engines, but the later ones from Whites had independent air pumps, which maintained a constant vacuum. Each of the ships was fitted with two twin-furnace through-tube boilers fired with coke. They were replaced by diesel engined ships in the 1960s.

List of Engine Builders

Ashton, Frost and Co, 52
Ashworth and Parker, 91
Baker, 71
Bever, Dorling and Co, 16
Booth and Co, 67
Bradley and Craven, 23
Bryan, Donkin and Co, 26, 87, 90
J. Clark and Co, 68
R. H. Daglish and Co, 14, 28, 42
Dodman and Co, 38
Easton, Amos and Co, 78
Easton, Amos and Sons, 75, 76, 81
Easton, Amos and Anderson, 79
Easton, Anderson and Goolden, 48
Easton and Anderson, 21, 24, 25, 82, 88
Fawcett, Preston and Co, 19
Finch and Co, 93
Fleming and Ferguson, 47, 94, 95
Fletcher and Co, 69
Gimson and Co, 15, 61
Goddard, Massey and Warner, 64
J. and H. Gwynne, 77
Galloway and Co, 102
Harvey and Co, 4, 5, 6, 7, 8, 18, 46, 57, 100
Hathorn, Davey and Co, 10, 11, 41
R. and W. Hawthorn, 2
Hughes and Lancaster, 66
Hutson and Son, 97
Hydraulic Engineering Co, 34
J. C. Kay and Co, 22

Kitson and Co, 9
The Lilleshall Co, 29, 36, 44, 62
C. Markham and Co, 30
Marshall, Sons and Co, 27, 33
Thomas Middleton, 86
R. Moreland and Co, 55
Neilson Bros, 17
F. Pearn and Co, 54
John Penn and Son, 35, 96
Pulsometer Engineering Co
    (Easton and Anderson Branch), 49
Rankin and Blackmore, 98
Rothwell and Co, 3
J. Simpson and Co, 12, 20, 39, 53
Sisson and Co, 91, 99
G. K. Stothert, 92
Stothert, Slaughter and Co, 59
Summers and Scott, 50?, 51?
Tangyes Ltd, 45, 65
G. Waller and Co, 89
Walker Bros, 101
J. Watt and Co, *Frontis*, 1, 40?, 58, 63
Westinghouse Co, 85
Willans and Robinson, 83, 84
J. S. White and Co, 103
Wood Bros, 60
Worthington Simpson and Co, 13, 37
Yarrow and Co, 43
Young and Co, 31, 32
Unknown, 56, 70

List of Engines Described

Water Supply

Cornish Engines

Frontis Whitacre	(J. Watt)	Birmingham Waterworks
1 Moors Gorse	(J. Watt)	South Staffordshire Water Board
2 Little Eaton	(R. and W. Hawthorn)	Derby Waterworks
3 Dudlow Lane	(Rothwell)	Liverpool Waterworks
4 Green Lane	(Harvey)	Liverpool Waterworks
5 Green Lane	(Harvey)	Liverpool Waterworks
6 Hammersmith	(Harvey)	West Middlesex Water Co
7 Hammersmith	(Harvey)	West Middlesex Water Co

Bull Engines

8 Campden Hill	(Harvey)	Metropolitan Water Board
9 Osgodby Station	(Kitson)	Scarborough Waterworks

Direct Acting (Non-Rotative) Pumps

10 Boughton	(Hathorn Davey)	Nottingham Waterworks
11 Cosford	(Hathorn Davey)	Wolverhampton Waterworks
12 Hammersmith	(J. Simpson)	Metropolitan Water Board
13 Marham	(Worthington Simpson)	Wisbech Waterworks

House-Built Single-Cylinder Rotative Beam Engines

14 Chelvey	(Daglish)	Bristol Waterworks
15 Hopwas	(Gimson)	South Staffordshire Water Board
16 Snarestone	(Bever Dorling)	Hinckley Waterworks

Woolf Compound Rotative Beam Engines

17 Cropston	(Neilson)	Leicester Waterworks
18 Crayford	(Harvey)	Metropolitan Water Board
19 Liscard	(Fawcett, Preston)	Wallasey Waterworks

Independent Rotative Beam Engines

20 Pembury	(J. Simpson)	Tunbridge Wells Waterworks
21 Brixton Hill	(Easton and Anderson)	Metropolitan Water Board
22 Dancers End	(J. C. Kay)	Chiltern Hills Water Co
23 Irton	(Bradley and Craven)	Scarborough Waterworks

Rotative Beam Engines, Gear Drive

24 Timsbury	(Easton and Anderson)	Southampton Waterworks
25 Timsbury	(Easton and Anderson)	Southampton Waterworks
26 Timsbury	(Bryan Donkin)	Southampton Waterworks

Horizontal Rotative Engines, Direct Drive

27 Luton	(Marshall)	Chatham and Rochester Waterworks
28 Eccleston Hill	(R. Daglish)	St Helens Waterworks
29 Prenton	(The Lilleshall Co)	West Cheshire Water Board
30 Linford	(C. Markham)	South Essex Water Board

Horizontal Engines, Gear Drive

31 Leamington	(Young)	Leamington Spa Waterworks
32 Leamington	(Young)	Leamington Spa Waterworks
33 Wilmslow	(Marshall)	Stockport Waterworks
34 Kemble	(Hydraulic Engineering Co)	Great Western Railway, Swindon Works
35 West Wickham	(John Penn)	Metropolitan Water Board
36 Southfleet	(The Lilleshall Co)	Metropolitan Water Board

| 37 | Eastbury | (Worthington Simpson) | Watford Waterworks |
| 38 | Marham | (Dodman) | Wisbech Waterworks |

Vertical Rotative Engines

39	Weston-super-Mare	(J. Simpson)	Weston-super-Mare Waterworks
40	Grimsby	(J. Watt)	Grimsby Waterworks
41	Aubrey Street	(Hathorn, Davey)	Liverpool Waterworks

Inverted Vertical Rotative Engines

42	Kirkby	(R. Daglish)	St Helens Waterworks
43	Wanstead	(Yarrow)	Metropolitan Water Board
44	Chelvey	(The Lilleshall Co)	Bristol Waterworks

Inverted Vertical Triple-Expansion Engines (Slide Valves)

45	Whatstandwell	(Tangyes)	Ilkeston and Heanor Waterworks
46	Ferry Lane	(Harvey)	Metropolitan Water Board
47	Falmer	(Fleming and Ferguson)	Brighton Waterworks

West Gloucestershire Water Company

48	Frampton Cotterell	(Easton, Anderson and Goolden)	West Glos Water Co
49	California	(Pulsometer Co, Easton and Anderson branch)	West Glos Water Co
50	Cowhorn Hill	(Summers and Scott)	West Glos Water Co
51	Cowhorn Hill	(Summers and Scott)	West Glos Water Co

Unusual Waterworks Engine Designs

52	Boughton	(Ashton, Frost)	Nottingham Waterworks
53	Hammersmith	(J. Simpson)	Metropolitan Water Board
54	Wilmslow	(F. Pearn)	Stockport Waterworks
55	Eastbourne	(R. Moreland)	Eastbourne Waterworks
56	Leith Road	(Unknown)	Gravesend Waterworks
57	Hammersmith	(Harvey)	Metropolitan Water Board

Sewage Disposal

Beam Engines

58	Chelsea	(J. Watt)	London Main Drainage
59	Deptford	(Stothert, Slaughter)	London Main Drainage
60	Markfield Road	(Wood Bros)	Tottenham Borough
61	Clay Mills	(Gimson)	Burton-on-Trent Council
62	Abbey Mills	(The Lilleshall Co)	Borough of West Ham

Other Types

63	West Dock	(J. Watt)	Hull Corporation
64	Poole Town	(Goddard, Massey & Warner)	Poole Corporation
65	Irthlingborough Road	(Tangyes)	Wellingborough Corporation
66	Eastbourne	(Hughes and Lancaster)	Eastbourne Corporation

Land Drainage

Beam Engines and Scoop Wheels

67	Misterton, Notts	(Booth)	Soss Drainage
68	Chear Fen, Cambs	(J. Clark)	Cottenham and Rampton Drainage Board
69	Halvergate	(Fletcher)	Sutton Marsh, Norfolk

Small Scoop Wheels

| 70 | The Sixteen Foot Bank | (Unknown) | Stonea District, Cambs |
| 71, 72 | Middleton Towers Estate | (R. S. Baker) | near Kings Lynn |

Scoop Wheels and Centrifugal Pumps

73	Pode Hole		Deeping Fen Drainage
74	Halvergate		Sutton Marsh Drainage
75, 76	Curry Moor	(Easton, Amos and Sons)	Curry Moor Drainage, Somerset

Gwynnes Pumps

| 77 | Misterton, Notts | (J. and H. Gwynne) | Soss Drainage |

Easton, Amos and Anderson's Drainage Engines

78	South Common, Lincs	(Easton, Amos)	West Butterwick Drainage
79	Westonzoyland	(Easton, Amos and Anderson)	Somerset Rivers Board Drainage
80	Aller Moor	(Easton, Amos and Anderson)	Somerset Rivers Board Drainage
81	Stan Moor	(Easton, Amos and Son)	Somerset Rivers Board Drainage
82	Waldersea, Norfolk	(Easton and Anderson)	Waldersea Fen

Electricity Supply

83	Herrison Hospital	(Willans and Robinson)	Charminster, Dorset
84	Avonbank Power Station	(Willans and Robinson)	Bristol
85	Shore Road	(Westinghouse)	Mersey Tunnel Railway, Birkenhead

Gas Supply

86	Old Kent Road, London	(Tho Middleton)	South Metropolitan Gas Works
87	Old Kent Road, London	(Bryan, Donkin)	South Metropolitan Gas Works
88	Winchester	(Easton and Anderson)	Winchester Gasworks

Gasworks Exhauster and Booster Drives

89	Stapleton Road	(G. Waller)	Bristol Gas Co
90	Bath	(Bryan, Donkin)	Bath Gas Works
91	Swindon	(Ashworth and Parker; Sisson)	Swindon Gas Works

The Smaller Ships

92	SS Tredegar	(G. K. Stothert)	Bristol Channel
93	TSS Bulldog	(E. Finch)	Bristol Docks
94, 95	TSS Alpha	(Fleming and Ferguson)	Salford/Mersey

Leisure Ships

96	PS Empress	(John Penn and Son)	Weymouth
97	PS Britannia	(Hutson)	Bristol Channel
98	PS Bristol Queen	(Rankin and Blackmore)	Bristol Channel
99	SS Henley	(Sisson)	Upper Thames

Tunnels, Docks and Floating Bridges

100	Severn Tunnel, Iron Shaft Pumps	(Harvey)	Sudbrook, Mon
101	Severn Tunnel, Ventilating Fan	(Walker)	Sudbrook, Mon
102	Newport Docks	(Galloway)	South Wales
103	Woolwich Free Ferry	(J. S. White)	River Thames